# The Vanishing Countryman

# The Vanishing Countryman

Edited by
G. E. Mingay

Routledge

First published 1989
by Routledge
11 New Fetter Lane, London EC4P 4EE



Typeset in 10/12 pt Times Linotron 202
by Input Typesetting Ltd, London
Printed in Great Britain by T.J. Press (Padstow) Cornwall



*British Library Cataloguing in Publication Data*

The Vanishing countryman
1. Great Britain. Rural regions. Social life, history
I. Mingay, G. E. (Gordon Edmund) 1923–
941′.009′734

ISBN 0–415–03429–9

# Contents

# The contributors

**W. A. Armstrong**, Professor of Economic and Social History, University of Kent at Canterbury

**C. W. Chalklin**, Reader in History, University of Reading

**B. A. Holderness**, Reader in Economic and Social History, University of East Anglia

**Alun Howkins**, Lecturer in Modern History, University of Sussex

**G. E. Mingay**, Emeritus Professor of Agrarian History, University of Kent at Canterbury

**Charles Phythian-Adams**, Head of the Department of English Local History, and Senior Lecturer, University of Leicester

**Michael Winstanley**, Lecturer in History, University of Lancaster

# Editor's note

Chapters 1 to 6 of this volume first appeared in *The Victorian Countryside* (ed. G. E. Mingay, Routledge & Kegan Paul, 1981); chapters 2 and 4 appear here in a revised form. The remaining four chapters, 7, 8, 9, and 10, appear here for the first time and have been specially commissioned for this volume.

# Introduction

One hundred and fifty years ago the rural community still accounted for about half the population of England – a surprisingly large proportion if the rapid growth of machine industry at that time, together with the irruption of Cobbett's detested wens, are taken into account. But from the mid nineteenth century the share of the countryside weakened rapidly, and already by 1880 the towns were able to boast twice as many people. In 1850, with the farm labour force at its peak, agriculture was still by far the largest single source of employment, and together with horticulture and forestry it supported over 2 million men and women. But even in 1850 this figure represented, as a proportion, only just over a fifth of the total labour force; and by the beginning of the twentieth century the 1.5 million then employed accounted for only one in every eleven of the employed population. The importance of agriculture continued to decline, though the fall in the absolute numbers is confused by the growing use made of part-time farmworkers. As a proportion of the total employment figures, agriculture steadily shrank to rank as only a minor employer, and came in the later decades of the twentieth century to make up only 2 per cent or less of the whole.

More recently, agriculture has come to be dominated by the owners of agricultural capital and employers of labour, the farmers. The farm-workers, still vital for their muscles and their knowledge, have in one sense declined relatively in importance as their numbers have fallen; on the other hand, those that remain have gained in standing as livestock and machinery have come to dominate farming systems, making skilled men key figures in the production processes. However, the third main element of the former tripartite structure, the landowners, have to a large extent disappeared as their estates have been broken up; and, as a result, well over half the farmland has come to be owned by its occupiers. Where landowners remain, their traditional role of leadership and management has declined with the modern restrictions on farm tenure and the growth of the tenants' economic and political independence.

The essays in this book are largely concerned with the first two elements, the farmers and their workpeople, but they also consider how the country way of life has changed beyond recognition over the past hundred years. The landowners, whose hegemony has been greatly undermined, not to say overthrown, during the same hundred years, have been over recent decades the subject of a number of specialized and well-known studies, and it has not been felt necessary or desirable to attempt to treat them afresh in the limited space of the present work. Nevertheless, it has to be remembered that up to the time of the First World War they were still a significant, even in some areas a dominant, force in determining the way in which farming was carried on, the use made of land, and the social life of a great many rural communities, though certainly not all.

Before 1914 the landowner-dominated rural community was very much the traditional ordered one, even if ominous cracks were appearing in the façade presented to the outside world. In 1895, almost twenty years before the First World War, and fifteen years after the onset of agricultural depression, Oscar Wilde scattered *The Importance of being Earnest* with references to the desirability of a fortune in the funds rather than one in an estate, and made Lady Bracknell utter the classic remark: 'Land gives one position and prevents one from keeping it up. That is all that can be said about land.'

When Wilde was writing the old rural order was transparently in decline, though only a generation earlier it seemed to have weathered the loss of the Corn Laws and the first major extension of the franchise and to be as strong as ever. The great territorial magnates stood at the apex of landed society, their wealth and pride exemplified by country palaces kept remote from the workaday world by vast rolling parks and deep belts of woodland. Together with the gentry and the other lesser satellites they controlled collectively much the greater part of the countryside. And their hold was not merely agricultural, for it extended to mines and iron-works, town properties, harbours, and inland means of transport. The more prudent landowners, indeed, were already following a policy of further diversifying their assets, buying stocks and shares, and showing an especial interest in railway securities at home and abroad. Nevertheless, it was the country estate that gave landowners their most cherished and characteristic source of income and power, the agglomeration of farms occupied by a respectable – and respectful – tenantry, with the land itself tilled by an army of labourers who, like the tenants, could for the most part lay no claim to possessing any stake in the land.

To serve the needs of landowners, farmers, and labourers there existed also a very numerous body of tradesmen and craftsmen. Many of them were highly specialized, though it was not uncommon for some trades or crafts to be combined, and for small farmers to engage in part-time supplementary occupations. There were corn, cheese, and hop factors, dealers in livestock, poultry, fruit, and seeds, maltsters, millers, brewers, and carriers. Large villages boasted considerable numbers of blacksmiths, wheelwrights, builders, carpenters, harness makers and saddlers, shoemakers, and tailors, as well as general shopkeepers, publicans, butchers, and travelling higglers. And these were occasionally the clients of professional men in country towns such as bankers, solicitors, doctors, estate agents, auctioneers, and surveyors. Some villages, it should not be forgotten, were partially or entirely industrialized, their inhabitants gaining a livelihood from a local mine, quarry, brickyard, tannery, or other works, or being involved in the manufacture, perhaps on a domestic basis, of boots and shoes, furniture, pottery, gloves, basketware, cordage, buttons, and a host of other small items. Where there were country houses a considerable permanent employment was generated for huntsmen, coachmen, grooms, foresters, gamekeepers, and gardeners, as well as droves of indoor servants.

Much of this declined or even disappeared towards the end of the nineteenth century and after, as farming became less prosperous, landowners sold or shut up country houses, and farmers relied more on family labour and employed fewer hired hands. Many country crafts were diminished or displaced by the decline of tillage in favour of grass, by the rise of urban factories, and the influx of imported substitutes for the products of the country works and cottage. Others, like the country coaches and carriers, felt the changes consequent on the spread of railways, and later the motor car; though the blacksmith, in particular, might keep going by turning to supplying and repairing farm machinery, or for as long as there were sufficient horses to be shod – and horses, though certainly diminishing, were still supplying a good deal of the power on farms right down to the Second World War. More recent trends, however, have revived some of these country towns as middle-class commuters and tourists have moved in. Antique shops, garages, and teashops have replaced the former smithies, corn dealers' establishments, and wheelwrights' shops, and new modern light industries have filled the vacuum left by the disappeared domestic crafts and small country works.

The estate, with its independent tenant farmers, labourers, outdoor and indoor servants, was not all pervasive. There were numerous communities where the land was more evenly divided, where small independent owner-occupiers survived, small builders were free to throw up cottages for the inhabitants, and the village folk in general could ignore squire, parson, and wealthy farmer alike. Even on estate lands the large capitalist tenant farmer, one of several hundred up to a few thousand acres, was untypical outside some areas of large-scale arable farming, as on the Lincolnshire wolds, the Norfolk sands, and Salisbury plain. The majority of farmers were much smaller, and indeed, two-thirds of all holdings of above five acres were below fifty acres in size.

The rural population was not as immobile, either in occupation or location, as is sometimes supposed. Census material reveals that many farmers, and indeed many other country dwellers, had not been born in the place in which they came to live, though a majority had not been born very far away. The growth of the rural population beyond local resources of employment, the lack of prospects and the desire to better oneself, marriage to a partner from another village or a town, the example of others who had prospered by moving away, perhaps abroad – all played some part in the movement of country people. They shifted from village to village, from village to a country town or to a major city, or they took the momentous step of emigrating. Among farmers and farm labourers the existing agricultural conditions, tenures, and custom might all be influential. The larger farmers, who held their land by lease, usually expected to move on when the term expired, while in the depression of the late nineteenth century many old tenants gave up, baffled by successive blows of poor seasons, low prices, foreign competition, and loss of capital. The opportunities which their departure created for renting cheaply run-down arable farms in eastern England proved attractive to Scots and west-country men, who saw an answer to low grain prices in economical systems of dairying for expanding urban markets. Such farmers, like many others, relied chiefly on family labour and hired few other hands, and this, together with the reduced labour demands of grass, contributed to the migration of farmworkers at this time.

The basis of the flight from the land was the expansion of town employment and the attractions of living where there was a wider range of jobs, where housing was better, living was cheap, and there was more to do in leisure hours. As people moved, and as schools, newspapers, and the culture of the towns impinged more closely on the village, so the old country lore was laid aside and eventually forgotten. Thomas Hardy was one of those who deplored the decline of the old tales and superstitions, the changes in habit and speech. Indeed, a surprisingly large body of traditional beliefs had until then survived the forces of change, but with the waning of the nineteenth century went also the wise men and women, the mystic chants and symbols, the love charms, the seasonal rituals, and strange folk remedies applied to all the common complaints.

In the new century the landowners moved more definitely into the ranks of the obsolete, and the independent farmer, owning his own land, came gradually to occupy most of the farmland, creating a situation which had certainly not existed for centuries or perhaps ever before. As tenancy declined, command of capital became of primary importance in agriculture. Of course, capital had always been important in some degree since having the means of working the land determined who could become tenants and of what number of acres; but now the farmer had increasingly to have not only his working capital but also that sunk into the land itself. The sum that was needed as farmland rose in price in the era of subsidized farming, and also as the size of the unit required for efficiency rose and requirements for buildings, stock, and equipment mounted, soon increased beyond the means of many would-be farmers. Such men, attracted to farming and increasingly likely to be trained for it, perforce had to look out for a scarce tenancy or settle for a job in one of the ancillary occupations, in the advisory services, perhaps, or in auctioneering or contracting. In consequence, over recent decades, the majority of new farmers have been farmer's sons exploiting inherited land and equipment and using borrowed working capital, and the larger operators have developed as kin-based partnerships.

The trends of government policy in regulating and subsidizing production have inevitably worked to the advantage of the larger farmer, since most of the official assistance has been geared to size of output. Farming has become more large scale and capital intensive, more scientific, and more technically complex, with the farmer essentially a capitalist businessman, well up in the latest technology and scientific jargon – very far removed from the somewhat rustic if hardheaded figure with stick and dog portrayed by an older generation of country writers. The changes have placed greater emphasis on education and training, assets in which former generations of farmers were distinctly lacking, and the modern agribusinessman is as likely to be seen at specialist seminars and technical conferences as at a Friday market or Saturday point-to-point. With modern farming has come, too, a revolution in the public perception of the farmer. He is no longer seen as the 'natural' countryman but rather as a menacing figure whose belief in the primacy of production means despoliation of the environment.

The farmworkers, like the farmers, have shrunk in numbers and changed in role. Machinery and the greater use of family labour on farms have reduced the demand for hired labour, though principally that of women and part-time workers. And, while the old traditional skills have become out-moded, new skills in the operation and servicing of machinery have in large degree taken their place. As the age-old movement of labour off the land has continued, the farmworker versed in the modern skills has acquired a scarcity value, and this, together with the establishment of wage regulation, has done much for farmworkers' wages and living standards. By the 1960s most cottages – the majority now of the tied variety – housed a television set, and farmers were beginning to complain of their workers' growing demand for the convenience of garages. But, by comparison with many other manual occupations, farmworkers were still near the bottom of the pay league, and there was, and remains, evidence of a significant level of poverty, particularly among families with children.

And in common with other low-paid groups, farmworkers lack the means of buying a house, and so are unable to join those many people who have found house ownership a valuable hedge against endemic inflation; and consequently they are unable also to gain the independence which escape from the tied cottage would mean.

Over the past thirty years more rapid change in the rural community has resulted from more extensive invasion by middle-class commuters, retired people, and town-based weekenders. Sociologists have argued that as a result villages have become mere extensions of the town, 'discontinuous suburbs'. The beginnings of the decline of a characteristic rural culture go back much further, of course, as indeed do the origins of the middle-class invasion, but the extent and implications of the recent developments have become apparent only in the past few decades. With the ending of the traditional dominance of squire and parson, the villagers acquired a greater equality and independence; but at the same time the changing social composition of many villages had made for a new inequality. The old inhabitants found themselves at odds with the newcomers' manners, habits, and ways of thinking, and found themselves unable to compete in living standards, housing, and transport. They are the ones who for the most part are confined to an alien council house estate situated inconveniently on the village edge, whose grown-up children have to move away since they are outbid for local cottages, and who are the most severely affected by the restriction of country bus services and closing of railway stations. And if they were to attend local evening classes, sports clubs, or drama or music societies – for which the majority, it must be said, show little enthusiasm – they would find these institutions largely dominated by their middle-class neighbours of superior attainments. Only a common interest in the village school might serve as a unifying influence: the Church, whose conscientious pastors in the past attempted to fill this role, has long fallen away and become neglected by the greater part of villagers, middle-class and working-class alike.

Thus the typical village of a large part of England in the 1980s - a former landlord's mansion converted into flats, a handful of well-to-do farmers whose production methods are anathema to the bulk of the community, a middle-class stratum who live in the village but are not of it, and those older-established local workers who form a disregarded minority relegated to the disdained fringe of late twentieth-century society. The origins of this situation go back at least for the hundred and more years covered by this book, even if the most rapid changes have occurred within the lifetime of its readers. One type of rural society has disappeared; another may as yet be only in its first phase of existence.

# 1

# The Victorian farmer

B. A. Holderness

Agriculture in nineteenth-century Britain was not a single industry but a combination of several, and the men (and women) who cultivated the soil were no less heterogeneous. The term 'farmer' was used to describe a rather residual occupational class. The holders of land who were not labourers, market gardeners, landowners, graziers, millers, innkeepers, and so forth, were 'farmers'. Thus the category included capitalists occupying 2,000 acres and smallholders with only 5 or 10 acres whose common interests, except perhaps as tithe- or rate-payers, were minimal. So indeed had long been the case, but earlier generations had made a useful if inexact distinction between 'yeoman' or 'husbandmen'. By 1851 farmers were readily differentiated only by the size of their holdings, but the diversity of agriculture was such that mere acreage was an inadequate indicator of social status. The standing, wealth, education, and life-style of farmers in different categories bore few marks of uniformity. At the top, farmers and their families almost, but not quite, merged with the professional classes and even with the lesser country gentry; while at the lowest level, the condition of the poorest farmers differed little from that of the better-off labourers and small village tradesmen. It is difficult, therefore, to discuss the 'place' of farmers in Victorian society. The larger capitalist farmers enjoyed most of the limelight and held nearly all the political and social authority vouchsafed to the Victorian farming community. The 'two-horse' occupier or the man without a team had about as much voice in public affairs as the labourer, and elicited less sympathy. Only in a few anachronistic districts such as the Pennine moorlands or the Isle of Axholme was there a kind of rough agrarian democracy in landholding, compounded of a peasant-like tenacity and an approximate equality of wealth.

Subsistence farming in the full sense scarcely existed by the second half of the nineteenth century, although it is probable that half or more of Victorian farmers consumed more of their own produce than they sold.

The capitalist farmers were in a small minority in the whole body of British agriculturists. Not all were men with 300 acres or more. The marginal productivity of the soil established different criteria for the optimum size of holdings in particular circumstances.[1] Some market gardeners were as affluent and commercially successful upon 50 acres or less as many cereal producers on 600 acres of scarp land. Moreover, as the century developed the bases upon which agricultural prosperity were founded changed significantly, and this shift not only transformed the practice of agriculture but wrought a number of important changes within the agricultural community. Victorians tended to understate this diversity and the process of change in favour of an ideal of progressive 'high farming', and in due course became entangled in the web of their own propaganda. The message of the depressed 1880s, received and understood by an intelligent minority in the eastern counties, was spelled out in the changed pattern of demand for farm holdings, for smaller farms were let more easily than large. The small family farmer, however, remained a rather shadowy figure in the 1890s, but the ability of his kind to survive both the mania for amalgamation and improvement and the new regime of 'low farming' after 1875 was demonstrated repeatedly during the century.

Victorian statistics indicate the numerical importance of the small farm. There were 19,400 holdings occupying 300 or more acres, chiefly in the eastern counties from East Lothian to Wiltshire, while the class of farms of 5–20 acres accounted for two-thirds of all the holdings above 5 acres returned in the enumeration of 1885. Even where, as Samuel Sidney suggested, the choice lay between 'large farms, much manure and numerous stock or no cultivation', farms of 50 acres or less were still to be found, clinging like moss to an ancient wall. In the Lincolnshire wolds, for example, there were 665 holdings of 5–50 acres and 216 holdings of 20–50 acres recorded in an official survey of 1890. The districts of greatest *morcellement* lay in the rich fens, especially in Axholme, and in the Pennine fringe regions of Lancashire and the West Riding near the growing industrial towns. The median holding in the pastoral west of Britain, where only Cheshire, Cumberland, and Somerset possessed a considerable number of farms in excess of 300 acres, was 65 acres in 1885, by comparison with 120 acres in the arable and stock-feeding districts of the east and south-east. The heavy clays, which were found everywhere but were characteristic of broad river vales or great plains, had a different profile, being neither fertile enough to support a thriving peasantry, nor sufficiently friable to promote the cause of high farming. The clays were typified by holdings of middle size (50–200 acres) devoted to stockraising or to an often rigidly traditional regime of mixed husbandry.[2]

Two obvious facts had a deep influence upon the structure of British agriculture. First, in spite of the pronounced dichotomy between large and small holdings, farm sizes in Britain, on average and in the lowest quartile of classification by acreage, were greater than anywhere else in western Europe. There were, for example, more 'agriculturists' assessed to Schedule B of the income tax in Ireland than in Great Britain.[3] Second, the overwhelming majority of farmers, except the gentlemen 'farmers-in-hand', were tenants. This was no less true of the far from negligible number of affluent owner-occupiers of farm land, nearly all of whom also rented holdings from other proprietors. The investment of savings by farmers, millers, maltsters, agricultural engineers, like that of the *nouveaux riches* of the towns, is a neglected theme, perhaps because it seldom affected the aggregate relationship between owner-occupied and *rentier* land in Britain before 1900. Every county could provide evidence of a few well-found ancient 'yeoman' families, like the Spurrells or Primroses in east Norfolk, or the Waights at Crawley in Hampshire, who survived or even prospered in Victorian times. In most regions of Britain, however, owners occupied only 10–15 per cent of their cultivable land after 1850.[4] The gains which may have been made by smaller landowners before 1815 were partly or wholly lost in the following twenty years. Tenancy, however, was a complicated institution. Not only were tenants often part-owners of their lands, but many held from more than one *rentier* proprietor. The large estates frowned upon the practice of hiring land from several owners, but less exalted proprietors had perforce to accept the divided loyalties of a tenantry, especially in difficult times such as the 1830s or 1880s. Many of the smallest estates depended upon the letting of land not equipped with buildings in order to maximize their net incomes from rent.

After 1850 there remained about a quarter of a million individuals admitted as farmers by the General Register Office, a slight fall from the numbers existing in 1831. The difficulty is to relate the data of the Census, the Agricultural Statistics, and the Inland Revenue. It is evident, although the numbers are uncertain, that Victorian agriculture supported a substantial body of part-time farmers. There were perhaps 100,000 holders of land whose occupation or social position was not that of farmer. On the fringe of the industrial districts, and in rural communities of complex social structure, smallholders and even 50-acre men could supplement their incomes by cartage, poultry higgling, or milk selling, and, conversely, many of the petty tradesmen and even some labourers also occupied plots of ground upon which they grew cereals and vegetables or kept nags, asses, cows, or swine. At the other extreme of the social spectrum, businessmen, solicitors, clergymen, as well as landowners, were often part-time farmers. Hobby-farming, as it was called, was quite fashionable in the nineteenth century. A few were like John Joseph Mechi, the London alderman who bought and cultivated Tiptree Hall estate in Essex, and spent many of his best years teaching and urging others how to farm profitably, but the majority had few such ambitions.[5] The variety of dual occupations in late nineteenth-century Britain is surprising after a century of more of industrialization and urban expansion. In districts of distinctively small holdings, Axholme, the southern fenlands, parts of Lancashire, and the West Riding, for example, it was apparently common for 'peasant' landholders to go out to work leaving wives and families to manage their holdings.

Although the numbers engaged in cultivating the soil of Britain were comparatively stable, a good deal of change occurred in categories of farm size. In the districts in which 'high farming' predominated by 1850 – the scarp-and-vale and former heathland landscapes of the east and south, from Strathmore to Dorset and Wiltshire – the work of arranging holdings of optimum size for maximum exploitation of cereal and livestock production had largely been done by the time of Repeal. Minor adjustments were made thereafter, although the trend towards amalgamation was less marked than before, and some of the bigger holdings were actually divided, even before the 1880s. Large-scale capitalist farming prevailed, chiefly because of the scale economies which could be achieved in light, dry soils, but most of the more enlightened estates, at least from Yorkshire southwards, operated a system of large, small, and middle-sized holdings. This encouraged some degree of upward social mobility among the tenants, and its social purpose was generally acknowledged by landlords like Lord Monson or the Earl of Yarborough. It was, however, also a consequence of the mixed soils which were to be found on the larger estates, for on the clayland arable the movement towards larger capitalist farms before 1850 had been slower and more hesitant. Even after a generation of heavy expenditure upon the clays, especially upon underdrainage, the obstacles to large-scale amalgamation remained. In the pastoral region of the north and west large farms of more than 300 acres were relatively less common. Even so it was in this area, especially in the north-west of England, that consolidation was most marked after 1820. The disappearance of the Pennine 'statesman' had been in train for a very long time before 1820–50, but the process was not complete even in the 1860s. In Cumberland, the Graham, Lowther, and Carlisle estates had undertaken to 'rationalize' their farm holdings by the middle of the century, as a result of which the number of tenants was much reduced. Thomas Farrell cited an example of one 300-acre farm which had been formed out of eleven former holdings.[6] In the more fertile plain districts of the west from Solway to Somerset the engrossment of farms in the interests of a successful mixed husbandry or of a more efficient dairy industry was active in the nineteenth century as a result of better superficial and sub-soil drainage and of the realignment of much of the farm layout of the region. Consolidation was not a constant theme of agricultural change in the nineteenth century: its progress and effects depended upon geographical and technical differences in the structure of British farming.

A larger question is raised if we turn from the number of farmers to consider the constancy of particular farming populations. Questions of inheritance, migration, recruitment, and upward (or downward) social mobility are difficult to answer from the surviving records. A study of particular populations at different periods suggests that there was a fair degree of volatility among British landholders.[7] A surname analysis in parts of East Anglia, for example, reveals that between two-fifths and two-thirds of the inhabitants described as farmers in each sample community had changed from about 1850 to 1890–1900, although many of the names recurred at different points in the same district during the later period. The census enumerations of 1851–71 tend to confirm this inconstancy. In a random sample of twelve English villages the number of farmers who were not born in the villages where they lived in 1851 averaged over two-fifths (40.8 per cent). Fewer than one quarter of these were not demonstrably the sons of farmers themselves, but there is evidence from a variety of sources that men were drawn into farming on their own account, from the ranks of labourers and servants and from the sons of clergy, tradesmen, auctioneers, or innkeepers. The opportunities for the farm labourer were slender but not completely closed, partly because even in the 1850s some landlords actively favoured ambitious, diligent, and frugal labourers or retainers in allocating small farms, but chiefly because in periods of low prices holdings were generally difficult to let, and many proprietors tended to encourage labourers with experience and small capital to take up tenancies. The successful were then able to progress up a ladder of promotion, and were not infrequently mentioned by agricultural writers of the time. In the farm prize competition of 1881, for example, one of the candidates, on a 70-acre holding, had himself risen from being a labourer. Even some of the bigger holdings in counties such as Lincolnshire or Northumberland were occupied by men who had begun as labourers during the last two decades of the century. However, it is difficult as yet to decide whether cases quoted in contemporary sources were representative of the general experience.

The pattern of geographical mobility is clearer. A minority of farmers migrated to seek better opportunities in other regions, and several of the most famous estates constantly received applications from far afield for vacant holdings. The best-known case it that of the Scots, from Ayrshire and Clydesdale, who descended upon East Anglia after 1880, to take farms which local men would not, or could not, occupy and, by their expertness in dairying, by their frugality and hard work, they rehabilitated many run-down farms.[8] This migration was exceptional only in its long range. Everywhere new tenants can be found moving in upon an alien landscape, and several estates, dispersed in different soils, actually encouraged good men to move from forward to backward districts.

The ubiquity of tenant farming in Britain signified that the most important of the relationships which existed between farmers and the rest of society was that which bound landowners and landholders. Tenant farming was not unique to Britain, but the symbiosis which developed between the two parties was regarded as the outstanding characteristic of British agricultural progress. Landlordism was not necessarily popular in the period, but more authoritative voices were raised in its defence than were deployed to attack it. The farmers apparently accepted the obligations and the benefits of the system in good grace. That is not to say that the history of landlord and tenant in Victorian times was uneventful or tranquil. On particular issues tension often ran high, but except in areas where ethnic or religious differences divided proprietors from occupants, in Ireland, Wales, the Western Highlands, or Shetland, disputes seldom flared up into conflict. Landlordism was an aspect of rural social order which, in some minds, was associated with traditions of hierarchy. The squire at the head of his little world of farmers and retainers, of shopkeepers and other clients, demanded a measure of deference or subservience which bore the colour, variously, of paternalism, feudalism, or tyranny (the point of view depended upon political judgement). But even squirearchal villages, graded in strata of social order, were seldom regimented in a recognized chain of command. The estate was a miniature social universe, orderly and patriarchal, but only a minority of British villagers, even of tenant farmers, lived near the pale of a great country house. Many farmers were tenants of more than one proprietor, and more than half of Victorian village communities were not directly under the influence of any great landowner. Indeed it was sometimes the farmers and village businessmen who formed the real elite of country society, not the landed gentry. As a political and social force the constituency of more affluent farmers had always been powerful enough to disturb the pattern of rural aristocracy. With their higher social standing and additional political power after 1830 capitalist farmers enjoyed a degree of public independence not matched since the apogee of the yeomanry before 1650. Thus while before 1900 there were villages which could be classified as feudal, and estates in which the will of the proprietor or the disposition of his agent were important in the ordering of social relations, obedience, except to the demands of good husbandry, was not necessarily a requisite of tenantry.

Deference is a different matter, for it appears that most farmers accepted the premise of the social superiority of the gentry and regarded deference as a natural element of good grace. On the other hand the social institution of tenant farming was regarded by most Victorians as a political instrument. Political disobedience could be punished by recourse to eviction, as was the case in Ireland and sometimes in Wales, but this was a sanction sparingly used. It caused too much disturbance and adverse publicity, and often resulted in the loss of an excellent batch of tenants. The furore which surrounded evictions in Merioneth after elections in the middle of the nineteenth century, and the even more famous scandal of that superlative farmer, George Hope of Fenton Barns, East Lothian, who had the temerity to stand for Parliament against his landlord's interest, and was subsequently dismissed when his lease ran out, were lessons which the majority of landlords heeded. The risk of dismissal for the average tenant farmer in Britain was less than the risk of bankruptcy or the risk of eviction for execrable husbandry.[9]

Victorian tenant farmers paid for access to the resource of land itself. Applications to occupy farm land generally outreached its supply, and in most agreements to hire a land a premium was levied, latently or quite openly, on its scarcity value. Offering holdings to tender was disapproved of by many agents and proprietors, but before the late 1870s it was a successful method of raising the level of rents in several instances. In the buoyant years from 1853 to 1873 most of the larger or better-managed estates enjoyed a waiting list of promising applicants, so that sitting tenants had to be circumspect in their dealings with the estate administration, particularly in chaffering over rent and expense. This sellers' market was seldom exploited to the full, and during the 1880s across the lowland zone of the island its metamorphosis to the tenants' advantage reinforced the message that good relations between owners and occupiers depended upon trust and fair dealing. Rents for the most part were determined by the soil, by the state of prices and the nature of production rather than by the simple willingness of tenants to pay a particular sum of money.

Interest on landlords' capital was a major element in rent. Interest, indeed, probably accounted for at least three-fifths of average rents in the middle and later nineteenth century.[10] The landlord's provision of buildings, drains, and fences was the principal justification in contemporary eyes for his control over farm land. Without his contribution the need to improve fixed equipment must have been met from other sources, from central government or from the banks, since the majority of occupants would have had insufficient funds from savings for such extensive capital formation as was achieved before 1914. Landlords were regarded by many farmers as desirable creditors, partly because the latter benefited directly from the fashion for conspicuous investment in agricultural estates, but chiefly because estate administrators were generally less exacting in their terms than institutional lenders. Between 1840 and 1880 the fashion for expensive buildings and drainage was such that few freeholders could have matched even by borrowing. Needless to say many tenants did not enjoy the almost perfect amenities offered by a Bedford, a Leicester, or a

Yarborough, but between 1850 and 1875 even the more backward estates began the process of modernizing their fixed equipment.

The formal distinction between landlord's and tenant's capital was seldom maintained in practice. The system was essentially flexible. Some landlords preferred low rents and little investment, leaving most of the improvement to the occupier; others shared the costs of fixed investment by providing materials while requiring their tenants to pay for the labour. There were, however, many variations of practice, most of which allowed the farmers some initiative or charged them for part of the expenditure. A legal impediment (*quidquid solo plantatur solo cedit* – that which is set upon the ground is appropriated to it), whereby a tenant's fixtures became the property of the landlord, did not hinder their efforts, because landlords seldom demanded their full rights at law, even against yearly tenants. This was important because with the decay of lease-holding, except in Norfolk and the Lothians, farmers' capital investment was largely unprotected. In due course systems of tenant right, by which outgoing tenants were recompensed for unexhausted improvements, developed outside the common law, especially in progressive regions such as Yorkshire and the east Midlands. Eventually the law was brought up to date, in 1875, 1883, and 1906, after which sitting tenants enjoyed enforceable rights against landlords or their successors. Yet without their contribution to estate capital formation the quantity and quality of fixed equipment would not have been less impressive. The landlords played the leading role before 1880, and in the last twenty years of the century, their need to retain sound tenants compelled them to shoulder almost all the burden of new investment. At the same time, landlords resumed another responsibility as creditors, that is, of supporting their farmers' business by remitting or abating rents, which had seldom been necessary in the 1850s and 1860s.[11]

Owners and occupiers were also enmeshed together in the observance of rules which governed the cultivation of the soil. The landlord's long-term guardianship of his estate led him to limit the liberty of the farmer to manage his holding as profit or short-run advantage might dictate. The long lease devised in the eighteenth century had laid down the ground rules of good husbandry, but fell into disfavour in the early nineteenth century. By 1850 the majority of British tenants were governed by yearly lettings. The effective difference, however, was insignificant, despite James Caird's opinion that the best husbandry depended upon the institution of long leases.[12] The essential point is that most annual tenants enjoyed security of tenure, even from one generation to another. For example, the stability of tenants on the Yarborough estates in Lincolnshire between 1815 and 1914 was almost exactly the same as on the Leicester estates in Norfolk, where lease-holding was maintained. Moreover, annual tenants were often bound by the same rules of management as lessees. Some husbandry clauses were regarded by many farmers as being too rigid, and several landlords certainly preferred standardized regimes of husbandry to arrangements to suit individual tenants.

Central as was the relationship between farmers and landlords, it was not all-consuming. The countryside was still densely populated, and the farmers' connections with the clergy, trades people, and professional advisers, and above all with the rural poor, were equally involved and significant in their daily lives. Most British rural parishes reached the peak of their population in 1851 or 1861, and although the decline before 1901 was often quite marked, rural society even in Edwardian England was occupationally and economically more diverse than has been the case since motor transport and commuter living altered the demographic balance of the countryside. The farmers were outnumbered by their labourers, and in many rural districts even by the rest of the gainfully employed population, but the largest proportion of the merchants and professional persons who still lived in the villages and small towns depended upon the prosperity of agriculture for their own livelihood. The wheelwrights, blacksmiths, millers, innkeepers, auctioneers, land agents, and solicitors of Britain's villages could not have flourished as they did without the business generated by the farmers. Early Victorian Britain in its rural setting was still essentially dependent upon handicrafts. Distributing and processing the produce of agriculture remained trades of major importance because so much of the goods and raw materials consumed in nineteenth-century Britain was still obtained from British farms. That the share supplied by foreign producers increased notably after 1850 is significant both for the development of domestic agriculture and for the changing structure of the agricultural service industries, but even in 1890 the connection between production and distribution or processing in British agriculture remained close. In 1850, however, farmers still sold their produce at local marts or to local dealers, and still also acquired their leanstock, seeds, horses, or tools within a similarly restricted horizon. Smiths were still called upon to make or alter implements or hand tools. It is not an accident that the majority of the agricultural engineers who set up in business on a larger scale between 1770 and 1870 had begun as blacksmiths.

Farming families were also important consumers of retail goods, and in every village there were grocers, shoemakers, and tailors, even, in a few drapers, wine merchants, and furniture brokers, whose business throve upon a clientele of farmers. Even country butchers depended upon slaughtering other men's fatstock for a good part of their income. The entirely self-sufficient farmer still existed in 1850 but he represented a dying species of peasant. During the period of fairly general prosperity from 1853 to 1873 much was achieved in rebuilding or modernizing country towns or large villages. New houses, shops, workshops, warehouses, and public buildings were erected in such profusion, especially in eastern Britain, that the outward appearance of many of our small towns is essentially still mid-Victorian. The process continued much more circumspectly after 1880. The local vitality of rural society was ebbing, for with the decline of population, the small world of traditional economic and social relations passed away. Decay, however, was matched by a trend towards business concentration, with economies of scale that broke down the old, disaggregated pattern of village economic connections. The common complaint that village shops were full of foreign products was interpreted as part of this malaise, but it should be taken equally to represent the first triumph of mass distribution in the British countryside.[13]

The emergence of large commercial firms handling milk, malt, bread-flour, and on the other side of the equation, producing and marketing fertilizers, feedstuffs, or agricultural machinery, affected the whole framework of rural society, modifying farmers' relations with their suppliers or merchants and undermining the livelihood of small country tradesmen. In milling, for example, two influences were at work from the 1860s, first the ever-increasing proportion of British bread-grains imported from overseas, which tended to concentrate the new industry in seaports near the chief centres of consumption; and second, the introduction into Britain of roller-grinding by entrepreneurs such as Joseph Rank, who not only stole a march upon competitors but were better able to exploit the supplies of hard Indian, American, and Canadian wheats. By 1900 flour milling in Britain was becoming concentrated in the mills of a few large firms, to which specialist corn merchants such as Quinton of Needham Market or Brooks of Manningtree were delivering great quantities of British grain. Furthermore, many middle-sized corn dealers, like R. & W. Paul of Ipswich, were also maltsters and feed compounders in quite a big way by the end of the century. A town like Ipswich, indeed, represented the new order of agricultural merchanting very well. It contained millers, maltsters, agricultural engineers, and milling engineers whose activities were very far removed from the millers and blacksmiths of Suffolk only a generation before.

The mechanization of British farming after 1860, and especially after 1880, was carried out through the agency of a few large firms and several smaller ones, whose interests were technically diverse and often international in scope. The small village craftsman survived the onslaught of British and American engineers, chiefly because he had adapted to repair or modify implements in the field. Whatever the farmers' views of machinery – and many resented it because of the high cost of mechanization relative to labour – machines and the network of distribution to supply and service them were part of the life of the larger commercial farmers by 1890. Moreover, the spread of expensive machinery gave rise to another rural occupation, that of the contractor who offered the use of cultivating or harvesting machines, particularly threshing boxes, and thereby spread the influence of mechanization further down the social scale.

The reorganization of the agricultural service industries did not coincide with a further bout of farm amalgamations. Big commercial farmers who had dominated the market for their produce and by their requirements in 1840–50, were less well placed by 1890–1900 to make bargains with their merchants. Co-operation was slow to take root in Britain, partly because of the prop of landlordism, but chiefly because there was little official encouragement for, and a good deal of individual apathy or opposition to, conjoint marketing. Only in 'residual' peasant areas, especially in the pastoral highland zone, did co-operative creameries become established before 1900, although the Irish example and Danish competition were constantly reiterated by enthusiasts such as Sir Horace Plunkett.[14] The problem was less serious than it might have been because the extension of rural railways and the growth of large firms in the processing industries tended to increase the demand for farm produce, even for cereals. In the particular case of potatoes, indeed, the decline which was observed about 1870 amid rising living standards was reversed by the efforts of entrepreneurs such as William Dennis of Kirton, Lincolnshire.

The influence of the railways upon the farmers was not completely beneficial. Rail transport did more than any other innovation of the age to create a national and international market for primary products. Without cheap rail transport in the New World the possibilities opened up by the free trade in grain and wool, and subsequently in meat and other products, would have been much more limited. Nevertheless, the economy of free trade which so aggrieved farmers was assisted by, not dependent upon, the innovations made in international transport, though home producers confronted with better or cheaper imports often confused cause and effect. Moreover, rail transport at home not only speeded up the process of rural depopulation, taking many of the better labourers and their families away from the land, but it also offered employment to countrymen. Again, railways were not the cause, merely the most effective means of townward migration. The positive contribution of railways within Britain was particularly marked in the reorganization of the distributive network of the agricultural economy. Rail transport made it possible for producers quite remote from centres of consumption to concentrate upon perishable commodities – liquid milk, vegetables, soft fruits – so that specialized areas of production – fruit farming in the Carse of Gowrie, around Wisbech, and in the Vale of Evesham, for instance, or market gardening in the Fylde or the fens – developed to supply the extensive new market for protein-based or green foodstuffs after 1860. Furthermore, the business of grazing, or livestock feeding and breeding, was changed with the expansion of rail carriage for live animals. Not only were the ancient droving trades modified by the railways, but new opportunities were opened up, as in the growth of the trade in Irish and even in Canadian leanstock after 1870. Farmers accepted the railways as a necessary adjunct of progress in Victorian Britain. There was little adverse criticism expressed about their economic role, for on balance railways probably improved and even cheapened the supply of farmers' goods; but in the case of particular railways local farmers were found in opposition to companies' construction plans, for permanent ways occupied useful agricultural land, and the farmers were as shortsighted as most in defending their vested interests.

The commercialization of agricultural marketing was obviously not complete by 1900, but the widening of the farmers' relations with merchants as well as with landlords was generally tempered by the influence of the proliferating army of representatives and local agents. The personal element was valued particularly because too few farmers kept adequate accounts and they seldom entered willingly into correspondence. The travelling salesman and the local dealer thus acted as a direct personal link in the lengthening chain of economic relationships between primary producer and distributor. Rural life retained much of the personal and parochial character which had distinguished village society for centuries. The social intercourse of farmers and labourers, tradesmen, clergy, and so forth was bounded by a restricted horizon as late as 1900. Contact with church, chapel, or school, interest in the affairs of the vestry or a local club, parochial alliances and antipathies, were the sum of the social activities of the majority of British agriculturists.

Victorian farmers were socially too diverse for a completely valid generalization to be made in a few words about their relationship with the church. Relations with the established church were sometimes very strained, especially in regions where the general population belonged to a religious denomination different from that of the official incumbent or minister.[15] This 'Irish' problem was rooted also in parts of Britain, in west Wales, in Shetland, and in the Catholic parts of the Highlands, and was aggravated by ethnic differences. In Wales, for instance, the Anglican clergy represented an alien elite whose demands affronted the Nonconformist traditions of the common farmers and others. Thus, those issues which everywhere were contentious, such as church rates and tithes, in Wales became matters to divide communities bitterly and even violently. Even in England many farmers were staunchly Methodist, and a few were Baptist or Congregationalist, which made their relationship with the Church of England uneasy if not antipathetic. In country districts the congregations of Wesleyan or other dissenting churches were frequently dominated by farmers whose authority resembled that exercised by their social peers in the parochial vestries. The wrangling over the control of education, over disestablishment, or parochial organization took place in rural England as much as in the towns, and in Scotland the dissensions which rent the kirk also disrupted the rural calm of many agricultural districts. Even within the Church of England the conflicts between High and Low, between Tractarians and Evangelicals, had much the same effect of suspicion and acrimony. Farmers, as a group, allegedly also held serious reservations about the social role of many of their parish priests. The gentrification of the English church, almost complete by 1850, in itself caused difficulties of communication beneath the sacramental level, however loyally and selflessly many country priests laboured to serve their flocks. The strong feelings which many parsons expressed about the plight of the rural poor alienated them from the farmers, so much so that William Johnston believed the parson's hat was more often to be seen hanging in a cottage than in a farmhouse.[16] The parson's championship of rural education, his agitation for, or provision of allotments, and at times his partisanship in such matters as the agricultural strike of 1872, were resented by many farmers, even indeed when they were loyal churchgoers. But the division among the farmers themselves on political, social, and above all on religious lines reduced the threat of an open breach based merely upon class interests. It may be true that most Anglican farmers belonged among the better-off, socially pretentious strata of village society, while the bulk of Methodists, however rich, were less easy to assimilate into the traditional pattern of the rural hierarchy, though as a generalization the latter is deficient in most of the finer shades of social distinction which were well understood in the Victorian village.

Anticlericalism, nevertheless, was not strongly marked in English or Scottish rural society. Most of the conflicts were over particular issues or personalities. Of the great issues, the duty of Nonconformists to pay church rates and the general obligation of tithes were the most important. Tithes posed a problem which the early Victorians purported to have solved, although since tithe charges were not exonerated before 1900 their payment continued to cause discontent in parts of rural England and Wales throughout the century. The Tithe Commutation Act of 1836 was intended to offer a standard system of levying tithes, to replace the variety of payments in kind or in money operated by local agreements. The commission established to discharge the provisions of the Act had largely completed its work by 1851. The plan was to commute tithe dues into an annual rent charge calculated on the basis of average prices over seven years. In operation it worked to the advantage of the farmers, for the base prices were low by the standards of the century, but in certain areas, notably in Wales, allegedly unfair assessments aggrieved the farmers and caused much unrest which culminated in the Rebecca Riots of 1842–3. But even in England the Tithe Commission could not smooth all anomalies. Before 1836 some parishes were already tithe free, partly as a result of special dispensation but chiefly as a consequence of parliamentary enclosure, since numerous villages had solved their tithe problems by allocating land in lieu of the charge. After 1850 the ancient complexity and diversity of the tithe system had been amended to such an extent that during the years of general prosperity before 1880 conflict fell away. At the end of the century, however, the relative burden of the tithe rent-charge increased, especially in eastern England, and a new agitation for reform or exoneration began. This culminated in the so-called 'tithe wars' of the 1920s and 1930s in counties such as Suffolk, from which emerged a plan for gradual exoneration after 1936. In Wales, where agriculture suffered less from the 'depression' of prices after 1873, the slow-burning antagonism between farmers and clergy or lay impropriators also flared up again towards the end of the century and led in its turn to disestablishment in 1914.

The parson's role as first or second gentleman in the Victorian village was not challenged by the majority of farmers. More often his position and his pretensions were ignored or disregarded. The average Victorian village had so many conflicts or suppressed discontents to divide or incense its people that the uneasy relations which existed between the parson and so many of his leading ratepayers were a matter of regret but not of surprise to contemporaries. Harmonious parishes did exist of course, but often only because the landlord had induced good feelings by force of his authority or by careful selection of the incumbents. Religion was too much a part of everyday life to be divorced from political or social differences.

The spread of rural education which was particularly active in the 1840s and 1870s owed little to the active support of farmers in general. Some farmers involved themselves in the building, equipping, and staffing of parochial schools, and the denominational battle over control of rural schools brought a number of farmers into the fray who were perhaps uninterested in the principles of education. Rural education, however, was dominated by parsons and ministers, by aristocrats and gentlemen, by exceptional men whose passion for progress and reform was unbounded, not by the generality of country rate-payers. The well-to-do farmers educated their own sons (and daughters at times) within the established network of schools, especially in the country grammar schools, while before the 1870s the poorest farmers often vouchsafed no book learning to their children whatsoever. Comparatively few farmers appear to have felt strongly that parish schools were an asset which would serve their interests. Indeed, for many the amenity of elementary education for their own kind was outweighed by the fact that educating the poor would give labourers false ideas of their worth and position in life. The majority of the farmers who ventured an opinion on the subject were scornful or even strongly antipathetic.

Although we know comparatively little about the social life of Victorian farmers, for few have left accounts of their entertainments and opinions, most certainly revelled in outdoor activities.[17] Hunting, shooting, and fishing were the most important pursuits, and despite the arrogation by the gentry of the best sporting amenities during the century, the increased leisure, social pretensions, and access to inexpensive sporting equipment of the more affluent farmers encouraged their participation in field sports. The Game Laws, which had reserved to gentlemen of substantial means the right to kill both wild and preserved game, were modified in 1831, but even then few tenant farmers expected to receive the privilege of shooting over their own holdings unless they were so far from the headquarters of the estate that their own game was not required by the gentry. In 1846 a select committee was informed that most of the bad feeling between landlords and tenants arose from game preservation, especially of rabbits and hares which damaged crops. It required another bout of prolonged agitation in the 1870s to produce the Ground Game Act of 1881, which authorized farmers to destroy rabbits and hares without their landlord's permission. Ground game, which caused so much strife between gamekeeper and poachers, between the rural poor and the country constabularies, was nevertheless of slight importance to many gentlemen, and long before the 1880s tenants were often allowed to take rabbits freely on their own holdings. Around mid-century the acquisition by farmers of sturdy and trustworthy guns, often of American manufacture, suggests that shooting, legally or illicitly, was a quite widespread pastime. A few of the most opulent were even invited to *battues* by enlightened landlords, such as the Earl of Yarborough. Exclusive rights attached to fishing, but since it was a less fashionable sport, licences to take fish in estate waters, except perhaps in the best trout or salmon streams, were more promiscuously granted.

Hunting was the great nineteenth-century field sport. It was, as some writers alleged, the cement which bound together countrymen of almost all ranks, one of the sturdiest props of the landed interest. Victorian foxhunting was well organized, dominated by coteries of landed gentry and precisely stratified in social ranking. The great hunts of the Midlands were fashionable and attracted many scions of the beau monde, but almost all hunts permitted or encouraged well-appointed farmers to ride with them. Throughout the hunting counties of England and Scotland, farmers and their peers, sometimes dressed properly in 'pink' but often in less aristocratic colours, turned out when they could. So many Midland and Yorkshire capitalist farmers bred hunters and carriage horses that a day's riding was an opportunity for self-advertisement as well as pleasure. But there was never a genuine freemasonry of the hunting field in the fashionable hunts, and hence the characteristic mid-Victorian development of farmers' hunt societies, open to gentry and others, but not dominated by them. Hunting retained its hold on even the foot-weary followers who could not afford to turn out mounted, the smaller farmers and labourers for whom drawing a local covert was great entertainment.

Field sports unified Victorian countrymen in a common pleasure. Other leisure activities are less easy to select, for farmers differed as much in personal inclinations as they did in wealth and social standing. Gentlemanly households, like that of John Pullett in *The Mill on the Floss* or even that of Farmer Boldwood, resembled those of the country gentry or *haute bourgeoisie* of the age. Such farmers' daughters were young ladies, and their sons were often educated to the law or the church. Card parties, musical entertainments, and refined reading were part of their lives, as they were elsewhere in the upper reaches of Victorian society. But there were other rich farmers, like Tennyson's 'Northern Farmer' or Richard Jefferies's John Hodson, of simpler, more vernacular stamp, who had neither polite manners nor elevated tastes. Most farmers belonged in this class, and their free-time activities resembled those of the labourers and village tradesmen. For some, no doubt, the public house was their Mecca, for others the chapel, and the narrowness of their lives was a matter of condescending observation among gentlemen, journalists, or large farmers. On the other hand, reading was certainly more extensive by the end of the century than it had been at the beginning. Country newspapers proliferated after 1855, and their attention to agriculture and local politics gave them a circulation which ran deeply into the rural population. Once upon a time, said J. C. Atkinson, the *Yorkshire Gazette* was passed from hand to hand among his farming parishioners until each copy fell to pieces.[18] By 1890 not only newspapers but magazines and broadsheets, some of them very lurid, could be found in even the remote farmhouse and cottage.

The newspaper merely reinforced what another custom among farmers had begun. The farmers' ordinary, offered at public houses in the country towns on market days, had long been used to make acquaintance and discuss prices and politics, and writers from Young onwards attended ordinaries to discover agricultural opinion. As an institution the ordinary was organized by the innkeeper and one or two leading farmers, so that it was easily converted into a vehicle of propaganda or persuasion. But market-day dinners were also enjoyable and companionable, and their influence in forming farmers' opinions deserves more attention from historians.

The social life of the Victorian farmer depended not a little upon the outward appearance of his life-style. Diet, dress, and above all housing distinguished farm families as much as any other social group. The house, in some sense, exemplified the family. Nineteenth-century farmhouses varied from the picturesque antique to the most modern brick boxes. A rage for rebuilding on many of the greatest estates from about 1840 to 1870 was transmitted to less affluent properties in due course, but even in the 1860s many leading estates still contained farmhouses of old-fashioned pattern and inferior materials. Vernacular styles had all but ceased after 1750, except in the remoter regions, in favour of brick or dressed stone with tile- or slate-hung roofs, but everywhere in Britain there survived weatherboard, wattle-and-daub, cob, black-and-white, and roughstone houses with thatched or shingled roofs which dated from an earlier period of extensive rebuilding between 1500 and 1700. Many of the smaller farmhouses were indistinguishable from cottages, cramped and inelegantly laid out. Indeed in the pastoral regions and the fenlands, where small farms still pre-dominated, the accommodation of farmers was a matter of quaint curiosity or downright condemnation.

This, however, was only part of the story. On the soils adaptable to Victorian high farming a rebuilding programme of major importance was accomplished before the 1850s, whereas in the clays the comparative poverty of the arable farms did not warrant expensive expenditure by farmers or landlords until the next generation. On the other hand, one noteworthy result of large-scale parliamentary enclosure, especially in the midland counties, was the construction of new farmsteads in the 'ring-fence'. That is to say, the old village-centre sites were abandoned in favour of a new, more efficient layout. Many Victorian farms bore names redolent of their epoch of construction. Quebec, America, Bunkers Hill, Salamanca, Talavera, Waterloo, Balaclava. We need go no distance to find a splendid example of Victorian agricultural brick or stonework. The farmhouse, 'commodious and convenient', surrounded by sturdy fold-yards, cart-houses, barns, and stables built to last several lifetimes, was an emblem not only of agricultural prosperity but also of proprietorial pride and ostentation. The farmhouses occupied by affluent tenants at Woburn, Holkham, or Brocklesby were fit for gentlemen. They typified an age in which well-to-do farmers had a large household of family and servants, and one in which material distinctions in standards of comfort and riches were no longer dependent upon the social order. Farmers were not gentlemen, but they could live like them. The reverse applied at the other extreme. There was no special cachet attached to being a farmer if you were poor.

It is probably best to assort British farmhouses into four categories from the standpoint of the late nineteenth century. First came the large, but diminishing, class of vernacular buildings, some of which were big, well appointed and comfortable, especially in the 'wood-pasture' regions. Many old farmhouses had been converted into cottage tenements by 1850, partly as a result of consolidated holdings and partly because of the relocation of farmsteads after enclosure. Others had been modernized, refaced, or reroofed to increase comfort or reduce fire risks. In 1850 the majority of British farmhouses were in this category; by 1890 the proportion had fallen significantly. Next came eighteenth- and early nineteenth-century houses in new materials, many built after enclosure but some merely the consequence of agricultural prosperity or landlordly concern in the years between 1690 and 1820. Third were the splendid and often extravagant houses built on the larger estates from 1830 to 1880, costing anything from £1,000 to £1,800. This fashion for 'high building' diminished after the 1870s as less money was available, but the visible remains of such work are still one of the most obvious features of Victorian farming life. Lastly, there was a large, undifferentiated class of more modest buildings, some constructed for small farms; others the work of particular owner-occupiers or tenants active in their own interest; yet others merely an expression of estate caution or poverty in providing new fixed equipment. Modest newly built farmhouses are everywhere in evidence from the period after 1830, as they were put up to replace dilapidated buildings or to satisfy the needs of medium-sized or small farmers. Such a farmhouse probably cost about £300 in 1850.[19]

By 1880 villages in which the farmyards still abutted onto the street in medieval style were few, as ring-fence farming became accepted as the standard of efficient management. But on clays or in backward areas generally, the impulse to remove the farmers out of the centre of parochial life was less potent, and old-fashioned villages such as Kirk Smeaton, Yorkshire, or Brooke, Norfolk, still retained several features of their ancient inheritance into the twentieth century.

The position of farmers in Victorian society at large changed as the prospects and contribution of agriculture changed. From its place at the head of the British economy in 1837, agriculture had declined sharply in relative importance by 1901. The contraction of agriculture owed something to the setbacks which afflicted arable husbandry after 1875, but it was chiefly the result of steady expansion in other sectors of the economy. The number of farmers indeed remained surprisingly stable after 1851 while the number of labourers declined significantly, a development which had a direct bearing upon the cultivation of the soil. The area of agricultural land had fallen by at least 1.5 million acres between 1830 and 1900, chiefly because of the spread of the built-up area. More important, however, was the much enlarged dependence upon foreign produce, even of products from temperate climates, which weakened the political and social significance of the farmers of Britain. Caird believed that about 21 per cent of our food supplies were imported in 1876, whereas forty years before the figure was probably only about a quarter of this. By 1909–13, 60 per cent of all British food consumption was imported.[20] Victorian agriculture continued to be efficient to the end, and the farmers were able to conduct a successful political lobby on many minor issues which concerned them. But on the great questions of protection and rural education the farmers' interest had been ignored or discounted in favour of larger political issues. This was not simply the outward sign of decay. It betokened a much more fundamental shift from a rural and agricultural to an urban and industrial society, in which the place of the farmer, in every sense, was bound to change. More remained of the lineaments of Victorian farming society in 1900 than the relative position of agriculture in the economy implies, for the complete absorption of agriculture into the framework of an industrial society was delayed until the Second World War; but the rural economy of the 1890s was no longer essentially hand-wrought and personal.

## Notes

1 Colin Clark, 1973, ch. 3.
2 Census of 1851, 1854; BPP 1886 [C.4848]; 1890, [C.6144]; Sidney, 1848, 4; Craigie, 1887; Grigg, 1963; Whetham, 1968; Phillips, 1969.
3 Caird, 1878, 300.
4 Sturmey, 1968, 283–6.
5 Mechi, 1845; 1859.
6 G. P. Jones, 1962; Farrell, 1974, 418.
7 From author's unpublished data.
8 McConnell, 1891; Lorrain Smith, 1932.
9 Davis, 1972; Olney, 1973; Crosby, 1977.
10 R. J. Thompson, 1968, 79–81; Bellerby, 1968; Colin Clark, 1973, ch. 6.
11 F. M. L. Thompson, 1963, chs 9–11; R. J. Thompson, 1968, 79–81; Holderness, 1972; Perren, 1970; Grigg, 1963; Perkins, 1975.
12 Caird, 1968, 27, 145, 347.
13 Graham, 1892; Haggard, 1906b; Sturt, 1912.
14 Digby and Gorst, 1957; Coleman, 1871; Pratt, 1906.
15 G. Kitson Clark, 1973; Evans, 1976, chs 6–7; 1975; Dunbabin, 1974; Johnston, 1851, II, 48; Ward, 1965.
16 Russell, 1965–7; Hurt 1961; 1968.
17 F. M. L. Thompson, 1963, 137–50.
18 Atkinson, 1891, 16.
19 Barley, 1961; Harvey, 1970; Peters, 1969; Denton, 1864.
20 Caird, 1878, 283; Flux,1930, 541.

# 2

# The workfolk

W. A. Armstrong

## The demographic background

The stages by which the medieval English peasantry disappeared, and the impact of successive phases of enclosure, continue to be debated. However, it is clear that the classic tripartite division of agrarian society (landlords, tenant-farmers, landless labourers) had already made its appearance by the early eighteenth century. Thereafter, we have to reckon with the factor of demographic growth which, at the national level, caused the population to triple between 1750 and 1850. The mechanisms of this increase have recently been reappraised in the monumental *Population History of England* by E. A. Wrigley and R. S. Schofield; some two-thirds of the increase after the mid eighteenth century is attributed to shifts in marriage and fertility levels, and one-third to favourable mortality changes.[1]

Within the national aggregate, the rural population grew considerably, though not so fast as that of the towns and industrializing districts. One calculation suggests an increase in sixteen primarily agricultural counties of 1.75 million (88 per cent) between 1750 and 1831, notwithstanding their loss by net migration of 0.75 million[2]. There are a variety of factors involved when accounting for this growth. First, as far as mortality is concerned, there was a distinct abatement in the incidence of epidemic disease and, in some parishes, a noticeable improvement in the survival rates of infants.[3] Second, in the early nineteenth century, land improvement schemes were deemed to have exerted a favourable influence, for example at Wisbech, Dunmow, Newhaven, Ongar, and in east Kent where the marshy land bordering the Isle of Thanet was effectively drained.[4] By the mid nineteenth century, low annual rates of mortality were a matter of common observation: 19 per 1,000 at risk (Cranbrook, Pately Bridge, Romney Marsh); 18 (Farnham, Liskeard); 17 (New Forest, Bideford, Hendon); 16 (Builth, Holsworthy); and even 15 (Glendale).[5] Although these figures relate to rural districts as a whole, farm labourers certainly shared in the favourable trend. According to statistics collected in the 1840s by the Manchester Unity of Oddfellows, only carpenters among twenty-five occupational categories had a higher expectation of life than rural labourers, who at the age of 20 could anticipate 45.3 further years; at 30, 30.7 years; at 40, 29.9; at 50, 22.2; and at 60, 15.8.[6] These advantages extended also to their offspring. By 1911, when the national infant mortality rate was 125 per 1,000 live births, the level in families of agricultural labourers stood at 97. This is not to suggest that there was no preventable wastage of life among them, for comparable rates for the offspring of solicitors and clergymen were 41 and 48 respectively; but they compared well with dock labourers (172), carters and carriers (147), bargees (161), and bricklayers' labourers (139).[7]

Trends in marriage and fertility must not, however, be overlooked. We need not dwell on the alleged nefarious influence of the Old Poor Law which was widely believed to encourage improvident unions and reckless breeding. These propositions, which persisted well after 1834, often depended upon highly coloured and untypical illustrations. For example, a contributor to the *Cornhill Magazine* in 1864 invited his readers to picture the labourer 'some fine day, before he is two and twenty, on his way from church, with his wife, who is only seventeen'.[8] But if contemporaries were often prejudiced about the explanations of rising fertility, it would appear that they were not mistaken about the trend. At the national level, Wrigley and Schofield suggest that between 1700–49 and 1800–49 the mean female age at marriage had come down from 26.2 to 23.4; while the proportion of persons never married fell from about a quarter in the 1670s and 1680s to some 4–8 per cent by the late eighteenth century.[9] Indications that these tendencies were at work in the rural as well as the urban districts are numerous. The mean marriage ages put forward by Wrigley and Schofield come, in fact, from a series of twelve family reconstitution studies which are based on village populations or at any rate on market towns[10]; but there is further support for falling marriage ages in a number of independent studies based on rural communities, such as Napton, Bidford-on-Avon (Warwickshire) and Powick (Worcestershire).[11] Furthermore, Anderson has shown that in 1861 the female age at marriage in agricultural registration districts which were dominated by labourers was, at 25.5, decidedly lower than that observed in districts where the labour force was more 'traditional' in character, i.e. heavily dependent on family labour and farm servants (27.3).[12]

In the face of such evidence the conclusion that shifting nuptiality patterns made a powerful contribution to rural population growth is inescapable, without age-specific marital fertility rates (i.e. rates of procreation *within* marriage) needing to move very far, if at all. A typical agricultural labourer married in the 1850s might be expected to father a family of five or six, and during the second half of the nineteenth century this average changed very little. Flora Thompson offers some interesting comments on the moral precepts which influenced procreation in late Victorian Oxfordshire. It was regarded as unseemly for grandmothers to bear children ('when the young 'uns begin 'tis time for the old 'uns to finish'); but conception prior to marriage was 'a common happening at the time and little thought of', and the control of births within marriage seems to have been frowned on in village circles. An admission of recourse to *coitus interruptus* met with the comment, 'Did you ever! Fancy begrudging a little child a bit o' food, the nasty greedy selfish hussy.'[13] At all events, according to evidence gathered in 1911, completed fertility among wives of agricultural labourers, standardized for marriage age, was 7 per cent above the all-class average for marriages taking place before 1851, and while the all-class average moved down by 21 per cent when the marriages of 1881–6 were compared, that of farm labourers was reduced by only 16 per cent: their fertility then stood at 14 per cent above the all-class average,[14] so that in a sense, relative to the rest of society, it was actually increasing.

If the analysis of the springs of rural population growth is a somewhat complicated matter, the tracing of its implications is scarcely less so. Recent research has emphasized that, over a very long period extending back into the seventeenth century, the rural population was not expanding so fast as that of the towns; and that the non-agrarian element tended to grow much faster than the number of agriculturists. Much of the rural population increase was channelled into domestic industry, still more into a wide range of trades and crafts, and a recent re-examination of the early nineteenth century censuses suggests that, as a whole, agriculture provided only about 100,000 new jobs in the whole period 1811–51.[15] These findings imply signal increases in agricultural productivity and they attest to the exceptional scale and speed of economic change in England on the eve of industrialization. But they are not incompatible with the presumption that, on the land, the ratio of landless labourers to farmers and landlords was increasing. Indeed, no other outcome could be expected, given that there were limits to the possibility of creating new holdings on hitherto uncultivated land and that a total restructuring of the basis of land tenure was unthinkable. Thus, in the village of Moreton Say (Shropshire), where the population doubled between 1680 and 1800, the number of labourers quadrupled. At Ash in Kent, which also experienced a doubling of population between 1705 and 1842, the number of holdings did not keep pace; in fact, they declined by a quarter, so again, raising the number and proportion of landless men.[16]

Nor are Wrigley's arguments inconsistent with the view that the forces of natural increase threatened, at times, to swell the number of potential labourers faster than the expansion of employment opportunities, so exposing men to the risk of seasonal or even year-round unemployment. It does not take much imagination to see that, had there been no migration outlets, and had non-agricultural occupations been less expansible than they were, the consequence of demographic growth would have been economically and socially catastrophic. In any case, as the agrarian disturbances of 1816, 1822, and 1831 suggest, the southern labourers at least were put under considerable pressure as a result of the failure of supply and demand of farmworkers to balance, and even after Victoria ascended the throne, their condition left much to be desired.

## Structure and composition

In 1851 the agricultural labour force reached its recorded zenith and, according to the census of that year, the ratio of labourers to farm occupiers was five to one. There was, of course, a good deal of regional variation reflecting prevailing patterns of farming and landholding, so that in Wales, where 72 per cent of all holdings were below 100 acres, the percentage of farmers with more than two labourers was only 17, and outdoor agricultural labourers accounted for only 36 per cent of all males engaged in agriculture. By contrast, in south-east England where 22 per cent of holdings were over 500 acres, 59 per cent had more than two labourers, who accounted for 82 per cent of all males in agriculture. At the national level, the 1851 census enumerated nearly a quarter of a million farmers and graziers (226,000 males, 23,000 females) as well as 112,000 male and 269,000 female relatives including farmers' wives. Their hired employees numbered 1,125,000 males and youths (chiefly outdoor labourers but including 102,000 described as servants) and 144,000 females, mostly indoor servants who numbered 99,000. After 1851 the number of employees began to fall.

Although there appears to have been a slight reversal of established trends during the Edwardian period, the pattern is otherwise consistent. There was a declining labour force in relation to farmers, whose numbers sank by only 10 per cent down to 1901, and thereafter rose by 2 per cent. This contraction was especially marked with respect to women, and it is convenient to begin with their contribution, with some reference also to children, who are excluded altogether from table 2.1.

By the early Victorian period, the systematic deployment of women, and to a lesser extent children, presented a society of increasing moral rectitude with a dilemma, noticed especially in two areas. One was localized, namely, the situation of the 'bondagers' of Northumberland, where the hinds were often required to provide an extra female labourer for farm work. These women, sometimes although not exclusively female relatives, excited admiration for their versatility and strength, and articulated as well a good deal of concern about their moral welfare and appar-

**Table 2.1** Employees in agriculture in England and Wales, 1851–1911

| | *Thousands* | | | | | | |
|---|---|---|---|---|---|---|---|
| | *1851* | *1861* | *1871* | *1881* | *1891* | *1901* | *1911* |
| Males | 1,124.5 | 1,114.0 | 939.6 | 850.6 | 776.5 | 637.5 | 674.4 |
| Females | 143.5 | 90.6 | 58.1 | 40.4 | 24.2 | 12.2 | 13.6 |
| Total | 1,268.0 | 1,204.6 | 997.7 | 891.0 | 800.7 | 649.7 | 688.0 |
| % change since previous census: | | | | | | | |
| (i) males | - | −9.3 | −15.7 | −9.5 | −8.7 | −17.9 | +5.8 |
| (ii) females | - | −36.9 | −35.9 | −29.8 | −40.1 | −49.6 | +11.5 |
| Ratio of employees to farmers and graziers | 5.08 | 4.82 | 4.00 | 3.97 | 3.58 | 2.89 | 3.00 |

Source: Taylor, 1955, 36–8.
Note: the table includes farm servants but seeks to exclude such categories as estate managers, gardeners, agricultural and forestry pupils, machine proprietors and attendants, woodmen, dealers, land proprietors, etc., as well as farmers' relatives. A few retired persons were included among the occupied before 1881.

ently feudal status. The other was a more widespread problem: the employment of women and children in public gangs, particularly in the eastern counties. Although there was no evidence of unusual ill health among them, the system was roundly condemned in the 1843 inquiry concerning women and children in agriculture on account of the loss of educational opportunities for children, 'imprudent' behaviour among the women, and the unbridled power it gave to gang-masters frequently described as 'low' and 'hard'. In 1867, following further inquiries by the Children's Employment Commission, a new Act sought to prohibit the employment of children under 8 and gangs of mixed sex, and to license gang-masters. This did not apply to the private gangs employed by farmers, which were in any case more numerous, and indeed, some of the twenty-two public gangs said to exist in Lincolnshire towns became private by the simple expedient of the farmer paying their wages directly.[17] The employment of juveniles was further curtailed by the Agricultural Children Act (1873) and the Education Act of 1876, but the passing of these legislative measures certainly did not mark an end to child labour, as the logbooks of many hundreds of schools show.[18] Furthermore, they were not directly concerned with the issue of women's work on the land. One explanation of the declining numbers of women in agriculture is seen not through legislation, but in terms of an increasing unwillingness to engage in field work. There is certainly a good deal of literary evidence to that effect. In the late 1860s Farmer Rollinson of Igborough (Norfolk) complained that he had been unable to get a woman worker in the last three years, while at nearby Salhouse it was remarked that the women 'did not care to come out' as their husbands' wages improved. At Felthorpe in the same county able-bodied women were said to prefer to walk three miles to the paper mill at Taverham, and by the 1880s Norfolk women were said to have 'passed out of the labour market altogether'. Likewise, in the Westhampnett Union (Sussex), female labour, 'once largely used', was by the 1860s rarely employed outside the hay and harvest seasons and at Slinfold there was 'scarcely one-tenth of the employment of female labour characteristic of twenty years before'.[19] In Northumberland, the bondage system which had relied upon a supply of unmarried females had almost died out by the 1890s: in Bedfordshire, straw-plaiting and lace-making provided a viable alternative in the 1860s but, said W. E. Bear thirty years later, 'now that the industries referred to are utterly unremunerative and nearly extinct the women still refrain from farm work'.[20]

An alternative approach to explaining the diminishing role of women has been put forward by Snell. To him, the Victorian statistics mark only the tail-end of a much more long-term tendency, going back into the eighteenth century and originating from competition within the farm labour force. This was especially marked in the cereal-growing counties of the south and east and was interrupted only briefly during the French Wars, and again for a few years in reaction to the pressures on family incomes arising from the introduction of the New Poor Law. His argument implies that women had little choice in the matter and that the direction of change was unmistakable long before Victorian moral sentiments concerning the proper role of women came into vogue, leaving them with only a limited range of options, which included domestic service (seen as a disguised form of underemployment), or securing husbands by getting pregnant.[21]

It is not easy to decide which of these explanations best fits the Victorian case. Indeed, it is entirely possible that the contribution of female workers (like that of children) declined a good deal more slowly than the census statistics first indicate. An interesting study based on some farm records in Gloucestershire, where according to the censuses female employment on farms declined by three-quarters in 1871–1901, suggests that the involvement of women was massively under-recorded: some, chiefly the wives of farm labourers, were working for as much as one-third or more of the farming year even in the late nineteenth century.[22] Even so, and for whatever reason, it is highly unlikely that the decline of female participation in agriculture is merely a statistical illusion, for when we turn to the male labour force, we can observe a corresponding trend towards decasualization.

This is not a process upon which the census statistics in table 2.1 can throw much light, partly because if the returns are presumed to show a man's main occupation, it will not necessarily be his only one. In practice, a variegated pattern of employment was facilitated by the widespread use of piece rates and short frequency weekly engagements. Moreover, since they were invariably taken in late March and early April, the censuses did not reflect either the size or the complexity of the labour force at peak seasons when the farmer's outlay on labour might well double or more.[23] To some extent these abnormal requirements were met by a redistribution of labour within the agricultural sector itself. Collins has identified one category of movement from grass and woodland pasture to arable areas, for example from the Vale of Gloucester and the cheese districts of north Wiltshire to the southern chalklands; from the Yorkshire dales to the East Riding; from the pastoral districts of Devon and Somerset to the Isle of Wight. Another category aimed to exploit the different timings and sequences of work between hill and vale, light and heavy land, and different farming systems, with a view to taking two or more harvests in a season. Yet a third category, and quantitatively the most significant, was between the 'small-farm subsistence and large-farm capitalist sectors of British agriculture'.[24] In the west Midlands this had formerly come chiefly from the hill counties of Wales, and in northern England entailed a flow of crofter folk from the counties of Argyl, Perth, and Ross and Cromarty, but by the Victorian period Ireland was a much more important source. The number of recorded immigrant 'harvesters' peaked in 1846–8 at the time of the Famine, and thereafter tended to decline, with a tendency for the Irish to fall back on those areas where wages were highest, such as Yorkshire, Lincolnshire, and the fen country.[25]

Traditionally, rural domestic industries such as hand-weaving had been another important source of seasonal labour, and through much of the Victorian period they remained so, employing polyglot armies of casual workers in towns, as Samuel has shown.[26] In the long run, and particularly after 1870, the importance of these 'wandering tribes' declined, for a number of reasons. These were in part technological, occasioned by increased use of the scythe and fagging hook in place of the sickle, and later the adoption of reaping machines which tended to lower the earning capacity of part-timers who were increasingly confined to the subordinate tasks of gathering and binding. Other factors included the greater regularity of employment in towns at wage levels which came to be higher than those which could be earned in the harvest field, and the gradual divorcing of an increasingly street-bred urban population from rural contacts.

However, the process of decasualisation was far from complete by the end of the century. In the Monmouth Union in 1893 many of the labourers were reckoned to rely for nine months in the year on other work, such as quarrying and mining, and in 1913 the Land Enquiry Committee guessed that some 100,000 farm workers (about one tenth of the total) could fairly be described as casuals.[27] This figure did not include those who continued to venture into the countryside to engage in market gardening, hop picking, and fruit picking, activities which were expanding their labour requirements through the Great Depression period as the arable acreage contracted. For those who in the 1890s worked in such disagreeable metropolitan industries as fur pulling, match making, and white lead manufacture, the attractions of hopping in Kent were obvious, and this particular seasonal influx remained in being until after the Second World War. But in agriculture proper there was no doubt about the trend. Agricultural commentators had long realized the value of the farmer's 'constant men' and advocated long engagements for specialist workers such as ploughmen.[28] Even in Wiltshire in the 1840s, where the overall situation of the agricultural labourer was poor, farmers would provide ten to twenty perches of land, ploughed and manured, together with a cottage and a good garden at 30–50s. rent per annum.[29] It is fair to assume that the easing by migration of the rural labour surplus would have had the effect of gradually increasing the regularity of employment of those who remained on the land, thereby increasing the farmer's reliance on his regular staff. In a sense, Wales offered the extreme case. Here the social distinctions between masters and men were very much less marked than in England and outdoor labourers were very hard to come by. In a primarily pastoral region requiring constant supervision for livestock, indoor service became the increasingly dominant form of hired labour in the later nineteenth century.[30]

## Wages and incomes

The labour surplus apparent in southern England during the early Victorian period seems to have favoured a tendency on the part of farmers 'to pay low wages in order to maximise employment for the men with poor-law settlements in the parish because the economic alternative of paying low wages for a small marginal product was unremunerative expenditure on poor relief'.[31] Against such a background cash wages showed no significant sign of advance before the middle of the century when the national average, according to Caird, was 9s. 6d. – no higher than in 1824. Thereafter it moved to 11s. 6d. in 1860–1; 12s. 5d. in 1867–71; 13s. 9d. in 1879–81; 13s. 4d. in 1892–3, and 14s. 5d. in 1898.[32] Mechanization seems to have had little impact on wages except in as far as it served to benefit a small number of operatives, such as the sixty 'formerly ordinary but intelligent farm labourers' employed by the Northumberland Steam Cultivating Company, whose wages had risen, in consequence, from 15s. a week to 20s. or 23s.[33] Nor did the organization of labour have a lasting impact. It is true that between 1871 and 1873 advances of 2–3s. a week were frequently achieved without industrial action, for example on the farms of Lord Braybrook at Audley End and Viscount Dillon at Ditchley in Oxfordshire, whilst in an essay on *The Dorsetshire Labourer* Thomas Hardy noted an average increase of some 3s. in this notably backward county.[34] But wages rose by at least as much in northern England, where there were very few trade unionists, and also in Wales, where it was said in 1892 that there was no trade unionism and nothing in the nature of a strike had ever been known.[35] Moreover, the waning agricultural unions proved powerless to resist effectively the loss of at least part of these increases as employers, beset by falling prices, retaliated in the later 1870s and 1880s. Rather, the long-term improvement was primarily due to changing conditions of demand and supply. Against a background of 'improved cultivation, more general and thorough management of root-crops, the extension of sheep farming, and winter feeding of stock' in the 1850s and 1860s, migration began to deplete the number of workers on offer so that by 1871 Dent considered over-supply to have become unusual.[36] As it gained further momentum migration frequently brought about a situation where the supply of labour was reckoned hardly equal to demand, as at Bryngwyn near Hereford where the counter-attraction of industrial employment in south Wales was keenly felt.[37]

However, nothing could be more fallacious than inferring earnings exclusively from the cash wages so far considered. Perquisites and allowances in kind played a large, if slowly diminishing part in the labourer's gross income. Notwithstanding the near-universal condemnation of the practice of supplying alcoholic beverages to field workers in the *Reports of the Special Assistant Poor Law Commissioners* (1843), beer continued to 'appear in the accounts of every farmer as an addition to his labour bill' in East Anglia,[38] and the practice remained very common in the southern, western, and south Midland counties. It was made illegal with the passage of the 1887 Truck Act, but continued nevertheless, as Rider Haggard was informed in the neighbourhood of Bridgewater in 1900.[39] Drink apart, the 'privileges' noted in 1867 in Devon and Dorset included the provision of a potato patch and cheap fuel or wheat although, as Hasbach points out, no farmer provided all together,[40] whilst in the 1893 Labour Commission there is evidence of the survival of various payments in kind.

If perquisites, and along with them extra harvest earnings, are reckoned as additional to cash wages, other factors worked in the opposite direction as potential deductions. One was simply loss of time in bad weather, although there was considerable variation of practice. From Wiltshire in 1893, it was reported that many farmers tried to keep their men on, wet or dry, in order to have a sufficient supply in the busy season. On the other hand, as Spencer observed after driving through the Essex villages of Latchingdon and Steeple on a wet day, 'most of the male occupants of cottages appeared to be at home or in the public house'.[41] Probably, as Clifford contended, the larger farmers were more ready in wet weather to pay their weekly men at any rate, if not those engaged in task-work.[42] The most comprehensive evidence on the subject, collected by the Land Enquiry Committee in 1913, suggested that time was lost by inclement weather in 47 per cent of parishes, ranging from 19 per cent in the north to 68 per cent in the south Midlands and eastern countries.[43] Loss of work through illness was another obvious way in which wages could melt away, and injury another, for sadly, accidents could give rise to atrocious conduct on the part of employers. In illustration Canon Girdlestone pointed to the case of an unfortunate carter who having saved a valuable team and waggon when they bolted, at the expense of having his ribs crushed, received from his ungrateful master neither wages, a visit, nor as much as a quart of milk for his children.[44]

So far as such factors could be taken into systematic account, Wilson Fox concluded that the earnings of ordinary labourers stood in a ratio of 119:100 to current weekly wages, although the variation was vast, ranging from 148 (Pewsey) to 106 (Uttoxeter and Wetherby).[45] Undoubtedly, as the Land Enquiry Committee maintained, labourers' earnings (as distinct from cash wages) were by far the more important figures,[46] and, whilst the earlier wage material does not lend itself to comparison, they appear to have moved on the lines indicated in table 2.2.

**Table 2.2** Agricultural labourers' earnings by region

| | *1867–70* | *1898* | *1907* |
|---|---|---|---|
| London area and home counties | 16s. 6d. | 18s. 5d. | 18s. 6½d. |
| South west | 12s. 5d. | 15s. 7d. | 16s. 5d. |
| Rural south east | 14s. 4½d. | 15s. 9d. | 16s. 10d. |
| South Wales | 12s. 7½d. | 17s. 0½d. | 18s. 2d. |
| Rural Wales and Herefordshire | 13s. 0d. | 16s. 1½d. | 17s. 8d. |
| Midlands | 14s. 1d. | 17s. 10d. | 18s. 4½d. |
| Lincolnshire, Rutland, Yorkshire (E. and N.R.) | 17s. 1d. | 18s. 0d. | 18s. 10d. |
| Lancashire, Cheshire, Yorkshire (W.R.) | 17s. 1d. | 18s. 8d. | 19s. 7d. |
| Cumberland and Westmorland | 18s. 6d. | 18s. 9d. | 19s. 2d. |
| Northumberland and Durham | 18s. 9d. | 20s. 5½d. | 21s. 5½d. |
| England and Wales, average of the regions | 13s. 9d | 16s. 0d. | 17s. 11d. |

Source: Hunt, 1973, 62–4.
Note: the unweighted average relates to fifty-four English and Welsh counties.

To what extent did such earnings support a rising standard of life? Prior to 1870 it is probable that any improvements were slight, and contingent upon a greater regularity of employment than had been obtainable in the 1830s and 1840s, coupled with an increasing solicitude for the welfare of their shrinking labour force on the part of more enlightened landlords and employers.[47] But with wages failing to fall commensurately with prices in the years that followed, there was a more noticeable advance. The falling cost of necessities was reflected in many ways. In 1893 Chapman noticed that on clothes lines good linen appeared instead of rags. The cottages contained a better standard of furniture, and every young man over 16 carried a watch. Butchers' carts called in the villages at least once a week, and not the least significant sign of progress was the appearance of lamps fuelled by cheap paraffin putting an end to the old habit of going to bed (or the public house) as early as seven or eight o'clock. At Chatteris a great many labourers took a weekly newspaper and patronized seaside trips, whilst there was a noticeable increase in the consumption of tobacco.[48] Although such comparisons were by no means entirely novel, in districts where small units prevailed, such as the Isle of Axholme, in Cumberland, and in Wales, the situation of the labourers was often favourably contrasted with that of the farmer himself.[49]

These impressions of a rising standard of comfort have to be qualified in several important respects. One critical factor, as Rowntree was so effectively to demonstrate in his study of York, was the family life cycle. Given the most favourable auspices (i.e. where there were adult children living at home) it was pointed out that gross family income could compare with that of a city clerk or a poor curate;[50] but as Rowntree's own excursion into rural sociology made clear, the presence of a large brood of younger children was a very significant factor in rural poverty,[51] and, as we have seen, the fertility of farm labourers was definitely on the high side. Secondly, even if we disregard significant local disparities at the district or parish level, there were well-marked regional variations in earnings, as is apparent from table 2.2. From at least the 1780s wages in the north of England had tended to pull away from those in the south. In 1850 Caird suggested that the difference was of the order of 37 per cent, and confidently ascribed this to 'the proximity of manufacturing and mining enterprise'.[52] There were differences of opinion among contemporaries about subsequent trends, as well as the importance to be attached to 'indulgencies' as a countervailing factor. In fact, Hunt's researches suggest that the percentage (of the British average wage) by which the maximum regional wage exceeded the minimum fell from 44 in 1867–70 to 28 in 1907. The important point to note is the persistence of variations, and his striking conclusion that 'wage differentials at any one time were as great as the overall improvement in wages between 1850 and 1914'.[53] Nor did the employment of women and children offset these differences, even in the first half of the Victorian period. Hunt concludes:

> Four shillings is probably a generous estimate of the average gross earnings of wives and children in the 1850's, and this was no more, and in many cases less, than the margin between farm labourers' wages in the north and south . . . [and] whatever residual compensations the rural south may have enjoyed at the beginning of the period were not enduring.[54]

## Characteristics of the agricultural labour force

With the southern labourer in mind, Hunt has argued elsewhere that 'sectors of the English agricultural labour force were living at a standard which, whilst adequate to sustain life, fell short of the level needed to ensure maximum labour efficiency', and that low wages were a consequence as well as a cause of low productivity.[55] This state of affairs was mitigated only slowly after 1870 as agricultural wages rose and regional variations became less marked. It is true, as his critics have pointed out, that as well as the varying quality of land, the relative amount of capital per worker would tell on labour productivity, and that his argument depends heavily on the citation of mainly impressionistic comments.[56] Yet it is striking how consistently the grain of this evidence runs in the same direction, that is, if one compares adversely the quality of labour in the south and east with that obtainable in Scotland, the north of England, and the north Midlands. In cases where southern labourers were transferred to the north it was often observed that they found difficulty in staying the pace. Thus in 1855 George Grey had attracted some two hundred southern labourers to dig drains in Northumberland by the prospect of earning 20–25s. a week by the piece. Within a short time only ten remained, and they never succeeded in making more than 15s., 'There was not a man among the whole importation that had legs and shoulders to compare with our lads of seventeen years of age.'[57] Caird, Clifford, Culley, Read, and Brodrick were just a few of the experts convinced of such disparities in labour efficiency, and Wilson Fox in 1906 was following a well-established tradition in ascribing them to the cumulative effects of 'generations of bad feeding' in the south.[58]

Poor nutrition, presumably, should be reflected in an above-average rate of sickness among farm workers, and there are conflicting impressions in the literature. On the one hand, Flora Thompson's recollections of Lark Rise in the 1880s are favourable: the doctor was rarely seen there and the general state of health was excellent owing to the 'open-air life and abundance of coarse but wholesome food'. At the other extreme we have Canon Girdlestone's view that the labourers of north Devon in the 1860s did not live, they merely 'didn't die'.[59] As we have seen, the rates of mortality prevailing among farm labourers were low by contemporary standards. Yet the Yorkshire doctor, Charles Thackrah, had emphasized in the 1830s that they were 'far less robust in figure than we would expect from the nature of their employ', and a few years later the published statistics of the Manchester Unity showed that although farm labourers enjoyed outstanding longevity, nevertheless they experienced 'an aggregate amount of sickness of 6.2 per cent more than the whole of the rural districts'. At age 20 the average was 4 days, 2 hours sickness; at age 30,

6 days, 5 hours; at 40, 8 days, 2 hours; at 50, nearly 14 days; at 60, nearly 27 days.[60]

The relationship between age and sickness revealed in these figures of Ratcliffe takes on further significance when the age structure of the agricultural labour force is examined, for as a result of age-selective migration it had come to be characterized by a comparatively high proportion of very young workers and an excess of the elderly, with a great dearth of men who were at once fully experienced and still in the prime of life, say, between 25 and 44. Notwithstanding comments like that from Zeal (Devon) – our young men have all gone, only old people and cripples left'[61] – W. C. Little contended that a comparison of the census returns for 1871 and 1891 gave no support to the prevailing impression of an ageing labour force; likewise the Registrar General was of the same opinion and succeeded in misleading Hasbach entirely.[62] In fact, a lateral rather than an historical comparison was more appropriate since the ageing process had been going on for many years. Such an approach reveals that in 1891 elderly workers (i.e. those aged 55 and over) were approximately three times as numerous, as a proportion, in the farm labour force than among railway employees or coal miners. A more broadly-based comparison is made in table 2.3.

**Table 2.3** Age composition of the male labour force aged 10 and upwards in England and Wales, 1891

| | | % aged | | | | | |
|---|---|---|---|---|---|---|---|
| | | *under 20* | *20–24* | *25–34* | *35–44* | *45–54* | *over 55* |
| A | Agricultural labourers, farm servants. shepherds | 28.0 | 11.9 | 16.8 | 12.7 | 11.9 | 18.6 |
| B | Remainder of male occupied population | 19.8 | 13.9 | 23.6 | 18.1 | 12.9 | 11.7 |
| C | % by which A exceeds B | +41.4 | −14.3 | −28.8 | −29.8 | −7.7 | +58.9 |

Source: BPP 1893–4 CVI [C. 7058]. *1891 Census, England and Wales. Ages, etc., Abstract.* Table 5, pp. x–xxc.

Youths loomed large as a proportion only because of the shortage of men in their twenties and thirties, and by this time even they were sometimes hard to come by. Thus in south-west Wales farmers had become reliant in part on the importation of lads from English reformatories, ragged schools, and industrial schools.[63] An Easingwold farmer remarked that they were 'quick in getting hold of machinery and interested in it, and in that respect better than the older men'; but, he added, they did not care to learn 'the old-fashioned arts'.[64] These were increasingly the province of the elderly, and the situation produced mixed effects. Often enough, and especially if they worked heavy land, men were 'very much bent' by their fifties.[65] Yet it was claimed in the 1870s, 'go where you will, you find old servants retained . . . sometimes receiving full wages, sometimes treated as "three-quarters" or "half" men, but hardly ever earning the wages paid them'. One of Clifford's Suffolk correspondents remarked: 'Neither I nor any decent farmer would turn a man off simply because he was old.'[66] From the employer's point of view sentimental considerations would often coincide with his interest in handling the lighter work of the farm cheaply, if with only modest efficiency. Taking this into account, Charles Booth remarked that 'in one way or another, effective working life is ten years longer in the country than in the town, or . . . is as seventy to sixty', and in a study of 262 rural parishes he found that 55 per cent of persons aged 65 and over could exist by their own earnings or means.[67]

Matters of the kind so far discussed do not exhaust the list of factors detrimental to labour efficiency. A very lengthy walk to work might result in labourers resting the moment the master's back was turned, as Farmer Norgate of Sprowston (Norfolk) complained in 1867.[68] Moreover, for many years the condition of the labourer, especially in the south, bred what Jefferies described as 'an oriental absence of aspiration'.[69] A Suffolk farmer recalled that when he asked labourers who had finished their stint of piece work by 1 p.m. to continue, the response was, 'No, master, we don't want no more money. We've as much as we care about! We'd rather go home and smoke a pipe.' Likewise in the west country the reluctance of labourers to forgo their cider allowances in order to secure a higher wage was much remarked upon in the 1860s and 1870s, and indicatively a meeting of eight men (with a total of fifty-two children) at Newent agreed, with one exception, that they preferred their cider allowance to 1s. 6d. extra wages, its cash equivalent.[70] It may be that the ensuing years saw a greater responsiveness to cash incentives but the process has yet to be traced in detail.

Many of these features served to encourage a stereotyped image of the farm labourers' bearing and address. According to Flora Thompson they detested nothing as much as being hurried,[71] and their sedate pace appears to have communicated itself to the young. Youths who, unlike their urban counterparts, had not been trained 'to appreciate the value of time' were criticized in characteristic terms by a speaker at the Framlingham Farmers' Club in 1867; if asked to fetch a rake a boy would 'open his mouth, turn his eyes on you and wheel on his heels with the precipitate motion of a Polar Bear'.[72] An article in the *Girls' Own Paper* in 1885 described the country lads of Dorset moving 'as though they have a heavy weight tied to each leg, so that it can only be moved by a heave of the whole body in the opposite direction'.[73] Such witticisms were heard more frequently with the passage of time: society as a whole, perhaps particularly the working classes, was prone to judge their own social and economic progress, albeit unconsciously, by the extent that they distanced themselves from the farm labourer's style and standard of life. There was point in their doing so. For notwithstanding a perceptible improvement in the agricultural labourer's condition during the Victorian era, it was pointed out in 1913 that in only five northern counties did his income reach the level necessary to avoid primary poverty,[74] and also – it would seem correctly - that he received a much smaller proportion of the wealth he helped to create than did his urban counterpart.[75]

## Notes

1 Wrigley and Schofield, 1981, 244.
2 Deane and Cole, 1962, 108. Their figures include the whole of Wales, classed for these purposes as a single county.
3 See, for example, Chambers, 1972, 97–106; Tranter, 1985, 45–6; Martin, 1976, 33–8; R. E. Jones, 1976, 313.
4 Flinn, 1965, 151–2.
5 Greenhow, 1858, 162–4.
6 Ratcliffe, 1850, 50. No doubt these rural labourers able to afford membership of the Manchester Unity were better off than the average, but this would also apply to other occupations to some extent. The data appears to cover some 17,000 rural labourers.
7 BPP 1912–13 XI [C. 6578], xli, xliii, 73–87.
8 Anon., *Cornhill Magazine*, 1864, 179. For a modern view criticizing the alleged impact of the Poor Laws see Huzel, 1980, 367–75.
9 Wrigley and Schofield, 1981, 255, 260.
10 Two of the parishes used were sizeable market towns (Banbury and Gainsborough), and others (e.g. Shepshed, Birstall) had significant industry. But at least half a dozen were primarily agricultural parishes.
11 Tranter, 1985, 50–1.
12 Anderson, 1976, 65.
13 Flora Thompson, 1939, 112, 142, 143.
14 Innes, 1938, 47.
15 Wrigley, 1985, *passim;* Wrigley, 1986, 336.
16 R. E. Jones, 1968, 9–10; information on Ash from Dr A. E. Newman.
17 BPP 1867–8 XVIII. Appendix pt I, 77.
18 See, for example, Horn, 1974, 56, 61, 71–8.
19 BPP 1867–8 XVII. Appendix pt I, 9, and pt II, 31, 36, 61, 77; Jessopp, 1887, 18.
20 BPP 1893–4 XXXV [C. 6894-III], 104; Agar, 1981, 31.
21 Snell, 1985, 45, 51, 56, 65, 66, 125, 348.
22 Miller, 1984, 145, 148.
23 For examples see Morgan, 1975, 39–40.
24 Collins, 1976, 43–5.
25 ibid., 50–1.
26 Samuel, 1972; Samuel 1975, 3–5.
27 BPP 1893–4 XXV [C. 6894-IV], 66; Land Enquiry Committee, 1913, I, 4.
28 See, for example, Wilson, 1851, III, 874.
29 Little, 1845, 177.
30 Howell, 1978, 93–4.
31 Digby, 1975, 79; Morton, 1868, 76.
32 Orwin and Felton, 1931, 233.

33 Dent, 1871, 348.
34 Horn, 1976, 132; Orwin and Whetham, 1964, 234.
35 Hasbach, 1966, 284; BPP 1893–4 XXXVI [C. 6894-XIV], 48.
36 Dent, 1871, 346–7; E. L. Jones, 1964, 328–9.
37 BPP 1893–4 XXXV [C. 6894-IV), 84.
38 Clifford, 1875, 117.
39 H. Rider Haggard, 1906, I, 249.
40 Hasbach, 1966, 337, 411.
41 BPP 1893–4 XXXV [C. 6894-V] 8, 76.
42 Clifford, 1875, 121.
43 Land Enquiry Committee, 1913, I, 21.
44 Quoted in F. G. Heath, 1874, 165.
45 BPP 1893–4 XXXVII [C. 6894-XXV], 84.
46 Land Enquiiry Committee, 1913, I, 4.
47 E. L. Jones, 1964, 331.
48 BPP 1893–4 XXXV [C. 6894-II], 45, 57, 83.
49 See, for example, BPP 1893–4 XXXVI [C. 6894-XIV], 172; Howell, 1978, 93; Dent, 1871, 361–3.
50 BPP 1893–4 XXXVII [C. 6894-XXV], 87.
51 Rowntree and Kendall, 1913, 33–4.
52 Caird, 1852, 511.
53 Hunt, 1973, 1, 58.
54 ibid., 121.
55 Hunt, 1967, 286.
56 Metcalf, 1969, 118; David, 1970, 504–5.
57 BPP 1867–8 XVII, Appendix pt I, 138.
58 Aronson, 1914, 63, quoting evidence to the Select Committee on the Housing of the Working Classes Amendment Bill, 1906.
59 Flora Thompson, 1939, 3, 141; quoted in F. G. Heath, 1874, 71.
60 Thackrah, 1832, 14; Ratcliffe, 1850, 50, 116.
61 BPP 1893–4 XXXV [C. 6894-II], 92.
62 BPP 1893–4 XXXVII [C. 6894-XXV], 33; Hasbach, 1966, 341.
63 BPP 1893–4 XXXVI [C. 6894-XIV], 9.
64 BPP 1893–4 XXXV [C. 6894-VI], 68.
65 R. Heath, 1893, 224, quoting Dr Batt of Witney.
66 Clifford, 1875, 120.
67 Booth, 1894, 321, 339. His figures exclude those in union workhouses.
68 BPP 1867–8 XVII, Appendix pt II, 29.
69 Jefferies, 1880, II, 78.
70 Clifford, 1875; 106; F. G. Heath, 1874, 88; BPP 1867–8 XVII, Appendix pt II, 133–4.
71 Flora Thompson, 1939, 46.
72 BPP 1867–8 XVII, Appendix pt I, 14.
73 See Kerr, 1968, 117.
74 Rowntree and Kendall, 1913, 31. Note, however, that they compared earnings in 1907 with prices current in 1912.
75 Aronson, 1914, 73. The statistics in Deane and Cole, 1962, 152, 166, confirm this impression. Thus in 1901 the share of wages and salaries in the total income generated in the agriculture, forestry, and fishing sector stood at 38.8 per cent against 48.1 in mining, manufacturing, and building, and 46.5 per cent in trade and transport.

# 3

# In the sweat of thy face: the labourer and work

Alun Howkins

> I think that the Tiller of the soil is the highest and oldest workman of all. No one can do without him and the product of his hands. The Gold miner cannot eat his gold, nor the Coal miner his coal, nor the Iron miner his Iron. All and every one is dependent upon the soil. He is the Father of all Workers.[1]

The farm workforce of the nineteenth century was far from being an undifferentiated mass of John Hodges. Although the national census of 1901 was the first to acknowledge the major divisions they certainly pre-date it, while even the divisions of 1901 – shepherds, horsemen, cowmen, and labourers – conceal the gradations of skill and prestige attached to these different jobs. On occasion these bland descriptions can be seriously misleading. Joseph Arch, for instance, was a champion thatcher and hedger and ditcher. At these crafts he earned enough to buy his own house and find employment throughout the south Midlands, although a known 'troublemaker'. Yet he would appear in the census as a labourer.[2] Similarly, oral evidence reveals an enormous variety of jobs and job descriptions. The father of one man I interviewed was by turns a labourer, thatcher, and quarry man, while another was a builder's labourer, farm labourer, poacher, marl digger, and fish hawker.[3] It would be sheer chance which of these jobs he was following on census day.

Further, there were regional distinctions. These produced a variety of localized categories of worker as a direct result of different types and patterns of farming. In Aberdeenshire the continued survival of the croft system, where married labourers held a small piece of land, created a peculiar intermediate stratum of labourer-farmers.[4] In south-west Wales up to the 1900s the men of the farms went to the coalfield in the winter months, returning in the summer to help pay off the families' labour debts.[5] In large areas of Northumberland and Durham the bondager system, whereby a labourer had to provide a woman worker (the bondager) to work with him, created another local category.[6]

Even beyond these local categories there were variations, meaningless perhaps to the outsider, but important in the village community. In the south-east, for example, it was usual for the horseman to be the 'superior' workman and therefore the older. When a boy went on a farm he went to 'learn a trade' which would stand him in good stead and increase his earning power as he grew older. Elsewhere, though, the situation was quite different. In Aberdeenshire, Cardiganshire, and the East Riding of Yorkshire, among other areas, the process was reversed.[7] Here, because of the living-in system and hiring by the year, the young men looked after the horses and the older, usually married workers, did the less skilled work. As David Jenkins has written of south-west Wales,

> A farm servant (*gwas*) was quite distinct from a farm labourer (*gweither*). Farm servants were unmarried and generally young men who were engaged to work with horses and lived in at the farm while labourers were usually married men who lived in their own houses. The general labouring work of the farm such as hedging, ditching, and drainage was specifically the work of the farm labourer. . . . The care of cattle . . . was for men the work of the lowest standing and accepted only as a last resort when nothing else was available.[8]

However, through all the regions basic divisions remained. Wilson Fox wrote in 1905:

> On farms of a sufficient size to admit of definite spheres of occupation being allotted, the work is organised, as far as possible under a system of sub-divided labour. With the exception of stewards, bailiffs, and foremen, the most responsible positions are those of the men in charge of the animals, and these are speaking generally, a higher paid class of farm servant than the 'ordinary labourer' and are usually on longer terms of engagement.[9]

In addition we can observe divisions within these categories. Firstly, within the category of men employed with animals it is important to divide them by the kind of animals they were working with. Secondly, it is necessary to divide the 'ordinary labourer' category into regular and casual. And, lastly, it is essential to note that the situation throughout the nineteenth century was not a static one.

The skills involved in farm work were many and various. Wilson Fox's division between those who worked with animals and those who did not seems to have been the primary one, yet quite how the division came about is difficult to determine. In most cases there seems to have been very little on-the-job training; as Jack Leeder said about the old horsemen he worked with, 'They weren't too good at [teaching] . . . you had to find out for yourself. They used to say, "Find out for yourself and you'll know how to do it." '[10] In many cases boys seem to have learnt from their fathers. A boy would go to work with his father from an early age to 'help out'. Arthur Amis, for instance, who was the cowman son of a cowman, did this,[11] as did Bert Hazell, whose father was a horseman.[12] Even this, though, could not guarantee that a boy would become a horseman or a cowman. There was always competition for jobs, and the individual's temperament was very important, particularly with horses. As Jack Leeder said, many boys were simply 'scared' of horses and had no control.[13]

The father-son situation, however, was almost universal in labourers' skills. A man who could thatch a rick, for instance, inevitably taught his son. Charles Leveridge started with his father when he was 11 years old: 'We used to have to go to pull the straw, before we left school, that was where I learnt my thatching.'[14] One suspects that this informal training in basic skills, and there were many, went right down the scale of farm work since all those interviewed spent a good deal of time, before and after school, in the holidays and playing truant, in the fields with their fathers.

The Victorian labourer, unless he or she was casually employed, was hired by the year. In most of north-eastern Scotland, Northumberland, Durham, Cumberland, Westmorland, Yorkshire, north Lancashire, and north Lincolnshire 'all classes' of workers were still hired and paid by the year in 1900.[15] In north Cambridgeshire and south Lincolnshire men hired by the year were mostly horsemen and shepherds. A similar situation existed in parts of the Midlands: mainly in Derbyshire, Shropshire, Staffordshire, Warwickshire, Leicestershire, Worcestershire, Oxfordshire, Berkshire, and Buckinghamshire. By 1900, however, even the yearly hiring of 'skilled' men went on 'to a much smaller extent than formerly'.[16] In the south-west the system 'was rapidly dying out', though it still continued in some parts of Monmouthshire and Herefordshire, and to a lesser extent in Hampshire, Dorset, Wiltshire, and Gloucestershire.[17]

Where men were hired by the year, they were engaged at a hiring fair ('feein' fair' in Scotland; 'sittings' in Yorkshire; 'mop' or 'statty' in the south). Here the worker stood in the streets on the day of the fair with a badge of calling in the lapel of the coat. Indoor and semi-indoor female servants were also hired by the year. A description of Bridlington hirings in 1895 catches much of the flavour of the fairs:

> Everyone tried to look smart; it is only right to say that. Many of the girls were brightly dressed, and only their speech betrayed them; but the lads still cling to the past in their sartorial get-up, which includes gaudy silk neckties and pearlies. . . . The waggoner has a bit of fancifully twisted cord in his cap, a bright flower . . . in his buttonhole, and his jacket not buttoned. . . . The proper fastening is two or three inches of brass chain, the better to display a capacious chest. Feathers on some

> of the bowler hats are suggestive of the fold yards, while the occasional flashes of bright colour in the feminine head-gear are suggestive of primitive Arcadia rather than the latest Paris fashions.[18]

Men and women hired in this way worked for a fixed wage plus board and lodging. This was the dominant pre-industrial pattern, though it had been gradually disappearing since the mid eighteenth century, especially in the economically most advanced wheat-growing areas of the south and east. As early as 1804 Arthur Young noted that the custom of living and feeding in the farmhouse was vanishing, while in Suffolk by 1813 he noted that 'the great mass of work in this county is done by the *piece*' [*sic*][19] which was the antithesis of yearly hiring. Where the system did continue the men usually lived in 'bothies', separate rooms over the stables, and the women in the house. On the smaller farms both sexes lived in the farmhouse itself.

The transitional form of employment between living-in and weekly labour was hiring by the year, when a man lived away from the farm but actually received his money weekly or fortnightly. This usually applied to men who worked with animals. In 1905 Wilson Fox noted that shepherds were almost inevitably employed by the year, as were stockmen, since 'it might put employers at great inconvenience if their shepherds or stockmen left them at short notice'.[20] This kind of contract was often extended to horsemen. The overwhelming weight of oral evidence suggests that horsemen (in Norfolk team-men) were employed by the year, and this is borne out by prosecutions under the Employer and Workmen's Act. Men employed to look after animals were graded. Each farm had a head horseman (in Norfolk head team-man, in the north head carter), who was the most skilled.[21] Similar grades existed for stockmen (in the south-east, cowmen) but not for shepherds, except on very large farms, as the shepherd usually worked alone with his page (boy).[22] This group of labourers was often housed in tied housing near the farm – the only substantial nineteenth-century group, other than those on estate villages, to live under this system.

Finally there were those groups hired by the week. Even here there was a sense that the contract of employment ran for a year. Wages books from Norfolk farms show that by the 1870s there was a regular core workforce of labourers, employed all the year round in many cases,[23] though often laid off in wet weather. This retention of the notional year's contract was related to the domestic economy of the labourer, especially the necessity of paying the rent once a year (usually at Michaelmas). In addition to these more or less regular workers there were those who were truly casual. The work of E. J. T. Collins and David Morgan has shown how the economic and organizational changes in agriculture in the early years of the nineteenth century created an enormous seasonal demand for casual workers. In the late 1860s this could mean a temporary increase in the workforce of 30–100 per cent.[24] Until the mid 1870s this extra workforce came largely from travelling and migrant workers. The Irish, the Scots, town-dwellers, gipsies, all once tramped the roads, following the same routes year after year. In some areas these migrants formed bands of skilled men and women who would travel from one farm to another taking harvests in turn. Such were the men of Blaxall in Suffolk described by George Ewart Evans, the travelling harvesters of Leafield in Oxfordshire or, in a different context, the travelling sheep-shearing gangs of Sussex.[25] As well as those who travelled, there were local workers who could be drawn upon for casual work. Crucial among these, until the mid 1870s again, were women and children. The practice of using the wives and children of labourers was at its most developed under the gang system. This was a system of subcontracting work to a gang-master, who then provided labour for particular tasks; it was widely used in Norfolk, Suffolk, Cambridgeshire, and north Lincolnshire between the mid 1820s and the 1870s.[26]

In the course of the nineteenth century all these regional variations in hiring, work, and classification were in a state of flux. In the south and east the dominant trend was away from a regular living-in workforce employed by the year towards a casual workforce living away from the farm, with a reduced number of regular and usually skilled workers. Thus at the end of the 1840s there seem to have been in Norfolk three casual to every two regular workers.[27] After the late nineteenth-century depression the pattern seems to have been reversed. As early as the 1860s there were complaints of labour shortage as rural depopulation bit and the supply of Irish casuals dried up after the Famine.[28] This really began to show at the end of the 1890s. By 1900 all observers were agreed that there was a serious labour shortage in rural areas, leading to much more regular employment. As Wilson Fox wrote in 1905: 'Generally speaking, since about the year 1896, ordinary farm labourers have been regularly employed. During this period farmers in all parts of the country have complained a scarcity of labour.'[29] The most eloquent testimony to this change is that by 1920 there were nationally five regular workers to every casual one.

The reasons for this are not far to seek. By the 1890s there had been a century-long drain on the over-populated rural areas, and even by the early 1870s some signs of crisis were apparent. However, the depression, coupled with some mechanization, softened the blow, and it was only with recovery in the 1890s that the full extent of the loss was clear. Additionally, in the south at least, women and children had been withdrawn effectively from regular involvement in the workforce by the gangs legislation of 1867 and the Education Acts. In the north the situation was different. The farms were smaller and had always competed with industry in terms of wages. This meant that a regular workforce, living-in, had been set as the pattern early in the nineteenth century: it was simply not possible to casualize the labourer when all could earn high regular wages nearby. Thus young men living-in, together with married men living out, but in regular work, was the northern pattern.

The actual work done by the farm worker varied according to skill and, crucially, season. This is best shown, in the first instance, diagrammatically. Figure 3.1 is based on the farming year in Norfolk at the turn of the nineteenth century, but with very little modification it could stand for the cereal-growing areas of the country for most of the century. The non-cereal modifications to this figure will be considered briefly at the end of this chapter.

The working year began and ended at Michaelmas or old Michaelmas. This was the end of harvest when the rent was paid and men changed jobs. As the daughter of a Norfolk team-man told me:

> In those days you see, farm workmen [at] Michaelmas, that was the 11th October in the country [old Michaelmas day], if they wanted another place they left the one they had and moved to another one . . . well then if they wanted to move again they just packed up and moved again.[30]

Another respondent said the roads were 'thick' with 'dicky carts' at Michaelmas.[31] Even here there was considerable regional variation. Michaelmas or old Michaelmas seems to have been the predominant time of year for moving in the arable areas, but elsewhere, especially in the north and Scotland, old May Day, Martinmas, or Whitsun were favoured.[32]

The horseman who came on the farm in the autumn faced a short period of intensive work, provided the weather was not hard:

> You'd start after harvest . . . what we called 'scaling', that's ploughing very fleet, which really cuts all the stubble and the weeds, harrow it . . . so it all pulls out, walk behind and lift the harrows . . . and of course that then all had to be shook about by hand, by fork, and then ploughed in.[33]

For the labourer, apart from forking behind the fleet ploughing, and muck carting, the main work of the winter was the root harvest. Early in the autumn the work, though hard, was pleasant enough:

> The method employed was this: you grasped the leaves of the mangold with the left hand. . . . You pulled the mangold out of the ground, swung it upwards, and at the right moment slipped your knife blade through the leaves where they joined the root. Then, if you had judged it correctly, the mangold flew into the cart and you were left with the

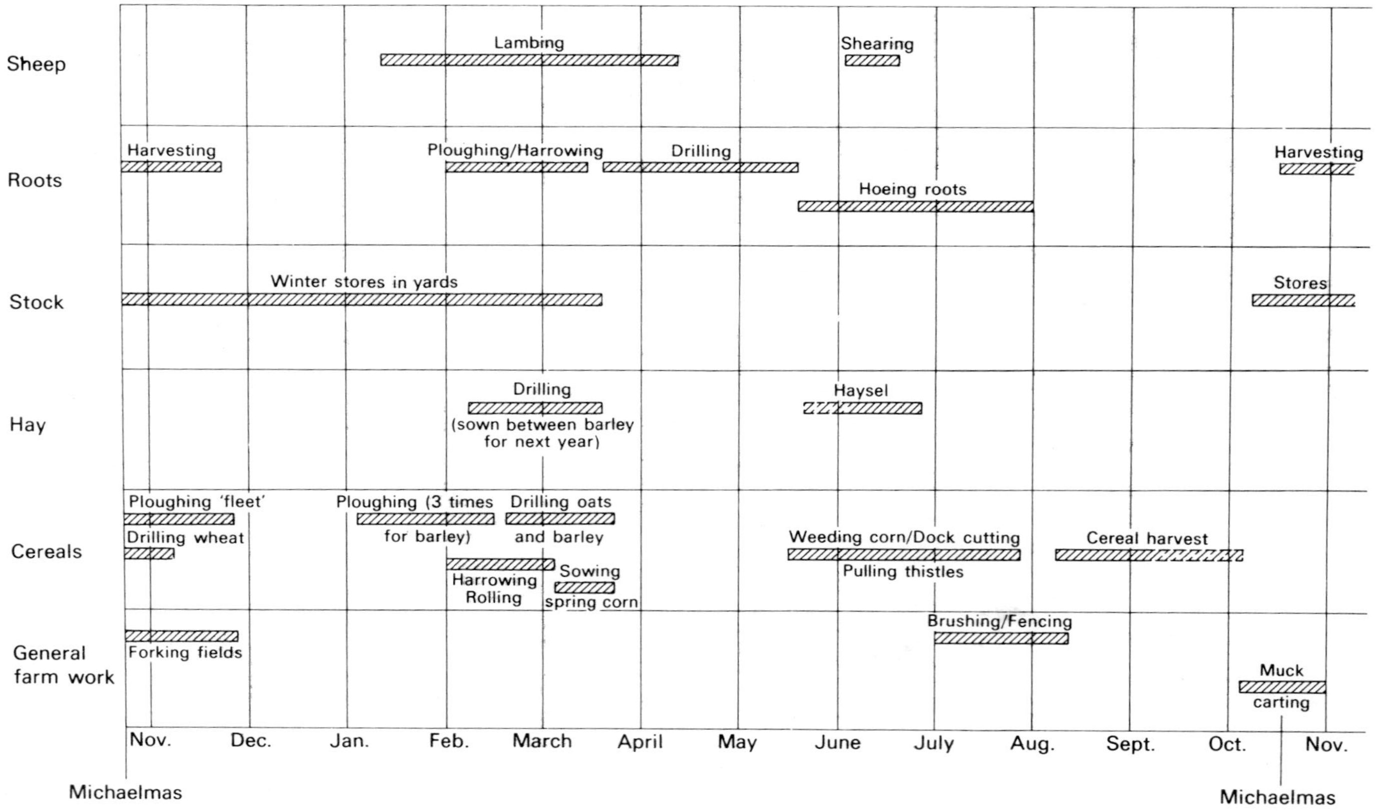

**Figure 3.1** The seasonal cycle of farmwork, Norfolk 1900–20, based on a four-course shift: hay/wheat/roots/oats

> leaves in your hand. You dropped them, and stooped to pull another. The whole process took the labourer one second.[34]

As the winter came on the process was less pleasant. Once the frost came the mangolds froze in the ground and getting them out became backbreaking work in which a misplaced swing of the knife could take a thumb or finger off. Along with the potato harvest, the root harvest in midwinter was the most disliked of farm work. If the weather held, ploughing fleet was followed quickly by ploughing for wheat and barley: 'they'd start and plough, plough as much as they could before Christmas'.[35] Ploughing for barley was a much longer process: 'they used to plough three times for barley . . . that was a big job, with all horses, and of course you couldn't rush'.[36]

Ploughing was the most skilled work of the most skilled men – the team-men or horsemen.[37] Under horse culture even a relatively small farm of 100–50 acres employed two or three team-men. A man who started work on a farm near Wyndham in Norfolk in 1919 talked about the organization of this side of the farm:

> You had five horsemen, five adults and a boy . . . [and] we all had six horse each, and then there would be another two or three looked after by someone else. They would be sub-divided so that we would run ten plough teams.[38]

The teams were rigidly organized. At the top was the head team-man: he was the most skilled, was paid more, and occasionally received a free cottage. He acted as a kind of foreman, 'keeping time and setting men off to work'.[39] This relationship was symbolized by the order of going out in the morning:

> When we were working in the plough teams [the head team-man] would take the lead, and nobody would dare to leave the yard until he'd got onto his horse and he'd got in front. He'd be followed by the second team-man and then subsequently down the line.[40]

Once in the fields he usually 'cut the field out'. Every forty yards across the field ('a forty-yard rig') the head team-man would cut one furrow, then turn it back. He was then followed by a less skilled man with a three-horse and double-furrow plough, who took his line from him. When the gap between the rigs narrowed the head team-man would finish the work.[41]

The skill required of a team-man was considerable and was surrounded by mystique. Even in the 1820s there were old horsemen who believed in and practised magic, mainly connected with the bones of the 'running toad'.[42] The younger generation were more sceptical. Jack Leeder, who started work in 1915, knew about the 'toad's bone' but 'didn't believe in it'. Nevertheless he did know recipes for making horses eat, making their coats shine, and dealing with cuts.[43] Team-men were very close about this kind of knowledge. All those I talked to, even the young ones, were extremely wary. Few learnt from their fathers: Jack Leeder's grandfather was 'a great horseman . . . but most of the things that he knew died with him'.[44]

To return to the yearly cycle. As winter set in farm work gradually came to a stop. Starting at the bottom men were laid off, although this practice was dying out by the 1900s. Nevertheless, the *Labour Gazette* regularly noted that numbers of labourers were out of work in Norfolk parishes.[45] Team-men and cowmen were seldom laid off in winter – their skills were too valuable:

> some of the labourers was laid off, but see that's where you got the benefit of it if you was team-man or yardman you got your full time in. But if you got a lot of bad weather, labouring chaps, they used to send them home.[46]

The period of laying off depended crucially on the state of the weather. In a really bad winter, for instance that of 1911–12, men could still be out of work at the end of March, having suffered three or even four months of unemployment. However, in most years farm work started up again soon after Christmas. Ploughing continued and drilling of spring wheat started as soon as possible. There was also threshing, although this was most usually done by travelling gangs hired by the threshing contractor. The men who worked the threshing tackle were a separate breed, men who for various reasons could not get or did not want a regular job. It was often the resort of a man dismissed for trade union activity, like Billy Dixon, who was sacked after leading a strike in a wood yard. He spent a year as 'second corn' on Bullimore's 'chining' crew from Bacton.[47] Another group who went threshing were fishermen. The enginemen often had experience of steam engines from the drifters, and many crew men 'did a couple of days threshing' in January and February when there was no fishing.[48] Threshing also provided work for the few true casuals as well as frozen-out farm workers.

In the middle of March, as the weather improved and the days grew longer, the workforce went on to summer time. The date of the change and the actual hours varied considerably. In Paston men went on to fifty-four hours on 1 March and wages went up by a shilling a week.[49] In Trunch they went on to sixty hours on 21 March for an extra shilling. Labourers worked all round the clock but team-men usually went two journeys, six in the morning until eleven, and one-thirty until six-thirty. The single journey was six until four.[50]

Also by March drilling was under way. This again was highly skilled work. If a man's ability could be judged by a straight furrow it could just as well be judged by the rows of green shoots as the corn came up. On a

Sunday it was a favourite pastime of team-men and labourers alike to walk around the parish and those nearby, gaining 'traveller' status and thus able to drink, but also examining the ploughing of their peers. A phrase like 'it looks as though a lot of bloody old chicken have been in there' could easily lead to a fight in a strange pub.[51] Drilling was usually done with a three-horse team, harnessed in a row, and led by a boy to get the accuracy needed.[52] Drilling followed a strict pattern. In many areas wheat went in in October if the weather was fair, though this was by no means universal. Then in February the rest of the wheat went in, then oats, then barley; in April, May, or even June the roots were drilled.

Between the end of drilling and the beginning of haysel (hay harvest) there was a period of slack:

> By the time you got your turnips and mangolds in there'd be quite a spell then, that'd be perhaps the slackest time on the farm. Then you'd do repairs and weeding of corn. There was a lot of dock digging them days and cutting thistles.[53]

At this time women and children appeared in the fields, weeding and picking stones. Stone-picking was paid by the piece – a penny a bushel[54] – as was weeding. As the roots came up they were thinned and hoed. This was done by the piece, usually by men working in gangs, the price negotiated by the head labourer, still called the 'lord' in some areas.[55]

At the beginning of June, again depending on the weather, haysel began. The hay crop was vital to the horse economy, and because of weather its gathering was fraught with some of the tensions which were so much a part of the later cereal harvest. For much of the century hay was cut by scythe, but by the 1900s mechanical mowers were becoming common. The haysel was perhaps the hardest of all harvests. Once the hay was cut it lay in the fields to dry, as the stacking of green hay could easily lead to spontaneous combustion. While in the fields the hay had to be turned by hand every couple of days, then raked into cocks and eventually loaded and stacked. There was seldom any extra payment for haysel and the men very rarely worked in gangs or by the piece. All this produces very different memories of haysel than the pictures evoked by the idyllic writings of many contemporaries. To the labourer the turning, raking, cocking, and loading in the heat of June, without the compensation of the extra money earned at harvest, was drudgery.[56]

Between haysel and cereal harvest there was another slight lull. In this hedges were cut, weeding of corn continued, and the fields were brushed. This was the laying of thorn bushes across the cut fields to stop poachers long-netting rabbits on them.[57] Then in mid-August cereal harvest began. As the corn yellowed the men would begin harvest bargaining. With the head team-man speaking for the horsemen, and the lord for the labourers, the rate for harvest was fixed on each farm. All watched their neighbours, as the first to agree usually set the price for the whole area, and even within a county there would be considerable variation.[58] When the price was fixed the men 'had a day hanging their scythes' at the blacksmiths. This involved sharpening and getting them set at the right angle in the shaft. Hanging was usually accompanied by drinking, with part of the cost borne by the farmer.[59]

The following day harvest began. Through the nineteenth century changes in technology can be observed – the change from the sickle to the bagging hook and scythe, and from the scythe to some form of mechanical reaper. Until well into the second half of the century the scythe predominated. Working in gangs of about twenty, the men cut in staggered line across the standing corn, taking their timing from the lord who stood at the head of the line: they stopped as he stopped and started as he started. Behind them a row of women scooped the corn into armfuls (sheaves or shooves) and tied them with bands of straw. In the stifling summer heat it was backbreaking work, but had a dignity and power about it that fixed this part of the labourers' work in the minds of those who saw it in its final years. One old man who had helped take the harvest with a scythe said: 'They were men in them days real men . . . they seemed more happy at work, they were continually whistling and singing. That's something I'm glad I've been able to be mixed up with, those old times.'[60]

By the mid 1890s the sight of a gang of mowers strung out across a field was becoming rare. Although a man still 'mowed round the edges' to clear a path for the mechanical reaper, the gang was gone. In its place the windmill-like blades of the sail reaper, and then the less poetic drum of blades of the reaper-binder, cut their way through the corn.

In the weeks of the cereal harvest the men battled against the weather. While it held they worked fifteen or sixteen hours a day. Rain would not only ruin the crop but crucially delay work. Since harvest was paid by the piece this could be disastrous. As one labourer said: 'The season ruled all, if you got a wet time you'd be about eight weeks, and your harvest [wage] was five pounds . . . and they were in debt [and] they'd drawn all their money . . . before they'd got half the corn in.'[61] But harvest did not end with the cutting of the corn. It was shooved, and when dry carted and stacked. Stacking was a skilled job. Wrong stacking or stacking too early could lead to rotting or burning. Crucially, it had to be thatched. Sometimes this was done by the head team-man, if he had the skill, more commonly by a local thatcher.[62]

When harvest ended there was a period of respite. The farmers would often go to sales while the men took their traditional harvest holiday. In the middle of the century, and up to the 1900s in some areas, they would 'cry largesse', visiting the market town and collecting pence from the tradesmen who dealt with their masters.[63] Later it was a trip by train to the county town or the sea to buy boots and clothes with the harvest earnings.[64] In some areas it was the start of the fair season. By the 1870s labourers flocked in their thousands by excursion train to fairs like Saint Giles in Oxford.[65] And then, after a brief glimpse of pleasure, the yearly round began again with moving place, ploughing fleet, and the bitter cold of the root harvest.

In sheep country, like Sussex, the pattern of seasonal work centred round the flock and took on a different rhythm, though few farms produced sheep alone and so the basic pattern of cereal production was still present. In the autumn in Hampshire sheep were turned out on to the stubble or folded with rape and turnips; in Sussex they went out on the downs.[66] Lambing, the shepherd's harvest, was between January and March, depending on the area and the mildness of the weather. During this time the shepherd lived out in the fields in a wheeled shepherd's hut, alone, except for his pages, for weeks on end. After lambing the shepherd had a period of ease: 'as a general rule, save for lambing and other busy times, a shepherd reckoned to finish his actual laborious work by dinner time. After that he studied your sheep.'[67] Dipping and shearing were in June: 'June was one of the busiest months in the sheep farmer's year. It brought none of the anxieties of lambing time, but in terms of sheer hard work, it stood out from all the other.'[68] A. G. Street summed it up well when he wrote:

> [Then] you will require swedes and kale for the flock. So your ploughs and harrows must follow them in May, June, and July as they feed on the rye, winter barley, and vetches, and then sow swedes and kale for winter. And so it went on, year after year, one continual hopeless striving to feed the flock. . . . Your life was ruled by them . . . the whole farm revolved round them. . . . They were a kind of Moloch, to which we were all sacrificed.[69]

Cattle reared for beef were frequently grazed in the summer in the upland areas, or even over in Ireland, and then taken south in the winter. In Norfolk cattle were 'stored' in yards and fed on roots throughout the winter, and sold in spring when prices were high and the fodder ran out. On better upland pastures the beasts were sold direct to the slaughterers. In dairying areas a more regular pattern of stock-tending emerged, in which the cereal cycle had practically no part and which produced a different hierarchy of workmen and workwomen. Where some cereals were grown dairymen were as separate and aloof as shepherds, as were the yardmen where beef cattle were reared. As Street said of the Hampshire dairymen, 'they ran their job without outside assistance. Give them the cattle and food, and they would do the rest.'[70]

Even where one system predominated there were almost endless variations of payments and work patterns. In Suffolk, harvest was taken by travelling gangs which were seldom seen in Norfolk, except in the south of the county. The huge variety of ways of paying for harvest work has been charted by Morgan,[71] but it extended to other areas too. In Sussex muck was sometimes carted by the piece and hay harvest taken by gangs,[72] while in south Oxfordshire no work was done by the piece at all, except the women's work of dock-pulling, weeding, and stone-picking, and the men received an extra 50*s*. at Michaelmas.[73] Twenty miles north, harvest and a whole range of other work were done by the piece.[74]

Nor were these processes constant, even within one area, although there was, in fact, surprisingly slow technical innovation in much of nineteenth-century agriculture. For decades the hay and cereal harvests were taken with bagging hook or scythe, stooked and loaded by hand, carted by horse, and stacked by hand. It was the 1870s before as much as two-thirds of all corn was cut and threshed by machine, and until the 1930s the great bulk of ploughing, harrowing, rolling, and carting was done by horses.

But changes came, changes which ultimately transformed the labourers' lot decisively, and for the better. The first was the steam- or horse-threshing machine. Appearing in the south in the early years of the century (though earlier in the north), it was well established by the 1850s.[75] The reaper and then the reaper-binder spread more slowly. The reaper became available during the third quarter of the century and the Appleby string reaper-binder from the early 1880s, but Rider Haggard did not buy his first reaper until as late as 1898,[76] and the reaper-binder was by no means totally established even in the 1920s.

It is common to attribute this slowness in adopting technical innovation to the conservatism of the farming community. Certainly many labourers were doubtful about change, preferring old ways. In the 1900s A. G. Street's 'improving' father came up against constant opposition, as did Henry Williamson in Norfolk in the 1930s. More recently many of the men interviewed by George Ewart Evans expressed grave misgivings about artificial fertilizers and the tractor.[77] Yet it is not that simple. The labourer was often mistrustful of change because he feared it might do him out of a job – as the threshing machine, the binder, and the reaper-binder certainly did. The master saw that good profits had been made under the old system, and was loth to risk high capital investment. Even the depression of the 1870s and 1880s did little to shake the faith of many.[78] Yet in the long term, among the labourers at least, few regretted the change: even the last generation of horsemen, those who started work in the early 1920s, welcomed the tractor in the end.

'In the sweat of thy face shalt thou eat bread', says chapter 4 of Genesis, and to the poorest of Victoria's subjects this represented literal truth as well as a biblical punishment. Farm work was unremitting toil in all weathers, from the sleet that accompanied winter ploughing to the burning sun of August. Harvest took a terrible toll. Every year boys 'riding holdya' fell from the horses and were crushed under the wagon wheels, and a scythe or an unguarded binder could cause a terrible wound which often went untreated and led to death.[79] Old men and women, driven by poverty to work in the fields at the busiest time of the year died of heatstroke, and in the end many were turned off the farm crippled with rheumatism or arthritis.[80] These things were as much a part of Victorian life as the skill of the thatcher or the dignity and strength of the mower and they are remembered as such by those who lived at the end of the horse economy. As one old team-man said, 'They were having their own way, the farmers was, they were putting us just anywhere . . . they never cared whether you lived or died.'[81]

## Notes

1 L. Haggard, 1935, 102–3.
2 Arch, 1898, 4, 57.
3 Interview with Herbert Neale, Paston, Norfolk; interview with Charles Leveridge, Carbrooke, Norfolk (tapes in author's possession).
4 Carter, 1976, 93.
5 D. Jenkins, 1971, 251.
6 BPP 1900 LXIII, 582–4.
7 D. Carter, 1976; Jenkins, 1971; Caunce, 1975, 45–53.
8 Jenkins, 1971.
9 BPP 1905 XCVIII, 357.
10 Interview with Jack Leeder, Happisburgh, Norfolk.
11 Interview with Arthur Amis, Trunch, Norfolk.
12 Interview with Bert Hazell, Wymondham, Norfolk.
13 Interview with Leeder.
14 Interview with Leveridge.
15 BPP 1900 LXIII, 582.
16 ibid., 587.
17 ibid., 587–8.
18 *Bridlington Gazette*, 16 November 1895.
19 Young, 1804, 484; Young, 1813, 223.
20 BPP 1905 XCVIII, 360.
21 For example, *Eastern Weekly Press*, 3 April 1903.
22 For example, Evans, 1956, 28–9; Copper, 1971, 65.
23 For example, Norfolk RO: Labour Book, Ditchingham-Hempnall area.
24 Collins, 1970; Morgan, 1975; Collins, 1976, 39.
25 Evans, 1956, 85; Morgan, 1975, 51; Copper, 1971, 116–17.
26 BPP 1867 XVII, *passim*.
27 Computed from censuses of 1831 and 1841 and other contemporary material.
28 Collins, 1976, 50.
29 BPP 1905 XCVIII, 354.
30 Interview with Mrs Moy, Yaxham, Norfolk.
31 Interview with Hazell.
32 BPP 1900 LXIII, 585–8.

33 Interview with Leeder.
34 Bell, 1930, 36.
35 Interview with Leeder.
36 Interview with Leeder; see Street, 1932, 40, for cross-ploughing for barley.
37 Evans, 1960, *passim*.
38 Interview with Hazell.
39 Interview with Leeder.
40 Interview with Hazell.
41 Interview with Leeder.
42 Evans, 1960, 260–71.
43 Interview with Leeder.
44 ibid.
45 *Labour Gazette*, February 1908, 55; November 1906, 340.
46 Interview with Charlie Barber, Great Fransham, Norfolk.
47 Interview with Billy Dixon, Trunch, Norfolk.
48 Interview with Sidney Watts,Happisburgh, Norfolk (tape at University of Essex); I am grateful to Trevor Lummis for this material.
49 Interview with Lee.
50 Interview with Dixon.
51 Interview with Amis.
52 Interview with Leeder.
53 ibid.
54 Interview with 'Butcher' Rayner, Swanton Morely, Norfolk.
55 Interview with Jack Sadler, Tichwell, Norfolk.
56 Interviews with Lee and Leeder.
57 Interview with Dixon.
58 Morgan, 1975, 45–53.
59 Interviews with Amis, Lee, Leeder, Dixon, and Neale.
60 Interview with Rayner.
61 ibid.
62 Interview with Leveridge.
63 Interview with Dixon.
64 Interview with Leeder.
65 Alexander, 1970, 28.
66 Street, 1932, 40; Copper, 1971, 65–6.
67 Street, 1932, 35.
68 Copper, 1971, 107.
69 Street, 1932, 40–1.
70 ibid., 36.
71 Morgan, 1975, 38–45.
72 Copper, 1971, 174–5.
73 Moreau, 1968, 66.
74 Morgan, 1975, 50–2.
75 Collins, 1972.
76 Rider Haggard, 1899, 274.
77 Street, 1932, 33; Williamson, 1941, 111; Evans, 1960, 17.
78 BPP 1895 XVII, 407.
79 *Eastern Weekly Press* 28 August 1902; 12 September 1908; 29 September 1900.
80 *Eastern Weekly Press*, 14 September 1907; 26 August 1905.
81 Interview with Dixon.

# 4

# The flight from the land

W. A. Armstrong

It is a fallacy to suppose that villagers were virtually immobile in distant times, for if their migratory horizons were usually very narrow, every serious inquiry so far conducted points to a degree of movement that is surprisingly high. Likewise the drift of population from the countryside was by no means a novelty brought about by the quickening pace of economic advance from the 1780s. On the contrary, towns of the pre-industrial era had always depended on a net inflow merely to maintain their numbers, and the migratory influx served to reinforce urban growth once the towns began to achieve self-generated natural increases, usually during the second half of the eighteenth century. However, the urban population did not clearly surpass that of the rural areas until after 1851, when they were approximately in balance and, absolutely if not relatively, the agricultural labour force of England and Wales stood at its zenith, 1.88 million.[1] By this time, with considerable improvements in the art of census-taking and the advent of civil registration in 1837, materials capable of giving a much clearer view of migratory movements were beginning to be collected. From the original census enumerators' returns it is easy to make a detailed analysis of the birthplace statements observed in any town, village, or hamlet, and it is not exceptional to find, for example, that three quarters of the adult population of an urban area were born elsewhere. As aggregated in the published census abstracts, the returns may also be employed to identify gross and net flows between counties. They do not, of course, throw any light on the intervening stages of migration, its timing, or the extent to which permanent and temporary moves are mixed together. On the side of vital statistics the Registrar-General's *Annual Reports* will yield natural increases calculated at the county and registration district levels, and, when compared with recorded inter-censal population changes, may be used to distinguish the components attributable to natural increase on the one hand and net migration on the other.[2] It is obvious that the migratory trends observed in a particular village do not necessarily represent those of the district or county of which it formed a part, and that alternative arrangements of the material may give rise to dissimilar, sometimes contentious interpretations.[3]

## The pattern of rural depopulation in the Victorian era

Summing up the basic statistics, which depend finally on the quantitative sources just mentioned, we may reach a number of broad conclusions.

1 Only three English and three Welsh counties (Cornwall, Huntingdon, and Rutland; Cardigan, Montgomery, and Radnor) recorded absolute decreases between 1841 and 1911. However, such calculations often obscure the factor of urban concentration, so that in Norfolk the aggregate census population of the three main urban centres (Norwich, Yarmouth, and King's Lynn) grew by 20.6 per cent and that of the remainder fell by 2 per cent.

2 More revealing are the statistics relating to registration districts (numbering altogether over six hundred), as analysed by Cairncross.[4] In fact the aggregate population of residual rural areas (that is, after subtracting the predominantly urban registration districts and those with extensive collieries) was actually somewhat greater in 1911, by 18.5 per cent in the north and 9.2 per cent in the south, than in 1841. Nevertheless their net losses by migration had been considerable, amounting to some 79 per cent of calculated natural increases in the north and 89 per cent in the south through the period as a whole, whilst, as table 4.1 shows, the drain was persistent.

**Table 4.1** Net losses by migration from rural residues in England and Wales 1841–1911 (in thousands)

| | *1841–51* | *1851–61* | *1861–71* | *1871–81* | *1881–91* | *1891–1901* | *1901–1911* |
|---|---|---|---|---|---|---|---|
| North | 159 | 229 | 254 | 263 | 349 | 237 | 152 |
| South | 284 | 513 | 430 | 574 | 496 | 423 | 142 |

Source: Cairncorss, 1953, 70
Note: the north includes for this purpose the counties of Herefs., Salop, Staffs., War., Worcs., Leics., Notts., Derby, Lancs., Yorks., Durham, Northumb., and Cumb., together with the whole of Wales.

3 While the phasing of the net outflow no doubt varied considerably among individual districts, table 4.1 indicates that it was most marked in the south as a whole in 1851–61 and 1871–81, but in the north in 1881–91. There is no obvious connection with the ebb and flow of agricultural prosperity, for as Cairncross has pointed out, the 'golden age' of the 1850s and 1860s showed a marked increase in migration and the recovery after 1900 coincided with its diminution, whilst in the intervening 'depressed' period the efflux from southern rural areas reached its peak and then subsided somewhat. Likewise Lawton has concluded that in general

neither quality of soil nor type of farming were important factors in explaining variations in the intensity or duration of the loss of population, save only from areas of arable land of the very highest quality where it was noticeably lower.[5]

4 Nevertheless, rural depopulation coincided with a substantial absolute decline in the agricultural labour force, which, while the number of farmers hardly changed, fell by some 23 per cent from its mid-century peak by 1911, and as a proportion of the total occupied labour force from 21.5 to 8.5 per cent.

## The situation of the agricultural labourer

The tendency for the number of employees per acre of land to decrease might appear to offer prima-facie support for a theory of labour displacement consequent on increasing mechanization, of the kind envisaged by Hasbach.[6] It is clear that an abundance of cheap labour had tended to militate against mechanical innovation in the years down to 1850 (with the sole, if notorious, exception of the threshing machine) and it is significant that a *locus classicus* of information on the agrarian labour force, the *Reports of the Special Assistant Poor Law Commissioners on Women and Children in Agriculture* (1843), mentions machinery only once. But mowing and reaping machines made their appearance in the 1850s and 1860s under the pressure of seasonal labour shortages, and something of their potential was illustrated by Caird's estimate that 80,000 would suffice to cut the entire harvest of Great Britain in ten days – a number, noted F. G. Heath in 1874, only double that actually used in the previous year.[7] The 1880s saw the advent of self-binding reapers, which, however, were less than satisfactory in dealing with heavy or laid crops. And more difficulties were encountered with the extension of mechanization to other activities. Although steam traction was applied to a variety of tasks, much resistance was met from farmers, not only on account of the expense but also, sometimes, on well-founded reservations about its efficiency. For example, steam-plough sets required a field of at least twenty acres and a terrain free from rocks. In the unpropitious circumstances of the 1880s, the steam-plough works at Dorchester was only saved by switching to the production of road-rollers, whilst well-informed observers on the heavy clays of Warwickshire thought that their use had 'somewhat declined' by 1893.[8]

It is true, reductions in the overall demand for labour were implied also by the shift towards pastoral activities during the 'Great Depression', and small farmers in particular were apt to seek immediate economies of outlay. Among the mountain of impressionistic evidence gathered for the Royal Commission on Labour (1893) the comment from Bishampton (Worcestershire) that not enough labour was employed so that several men had been 'driven away,' was by no means an isolated expression of view. Yet the reporter on four northern counties thought that implements were 'often used on account of the dearth of labour, not as a means to enable it to be dispensed with', instancing an encounter with a farmer who purposed a visit to York that very day to acquire a new self-binder 'because he could not get hands enough to do the tying behind his old reaper, except at a ruinous price'. Surveying the whole body of evidence in his general report, W. C. Little was inclined to think that the reduction of working staffs was a consequence and not the cause of migration.[9] Indeed, many witnesses remarked that their brighter young workers had a positive interest in and inclination to work with machines, impressed perhaps by the hauteur and high wages of the emerging class of 'agricultural machinery operatives', numerically insignificant though it may have been. In general it seems that where reductions in staffing occurred, they usually reflected either a disinclination to fill all the vacancies arising from death, old age, and dismissals, or reduced demand for 'catch' or casual labour.[10] The social implications of the recourse to machinery were accordingly not inconsiderable, particularly in regard to harvest earnings. It was reported from Wiltshire, that owing to the 'perfection' of machinery, the harvest could be secured in as many days as it used to take weeks, given good weather; and that earnings instead of being from £6 to £8, were considered good at £3.[11] Yet all in all mechanization seems to have had little impact in forcing involuntary migration upon the regular labourers, or even in forcing down wages. If it had any such inexorable tendencies they were clearly offset by other factors.

By far the most easily quantifiable contrasts between agricultural and other employment concerned wages and hours of work. Regional variations in farm workers' wages are discussed elsewhere, and here it is necessary to note only that the ratio between agricultural and industrial wage rates remained approximately constant at just under 50 per cent throughout the period.[12] What is more, it is now established beyond any reasonable doubt that gains in money wages arising from migration are unlikely to have been offset by increased living costs. It is true that urban rents were higher, the usual level for three rooms in large provincial towns at about the end of the century being roughly double, and in London perhaps treble, what the rural labourer would normally pay for his cottage, i.e. 1*s*. 6*d*.–2*s*. per week. Otherwise the cost of living was actually lower in the towns as a consequence of the gradual urbanization of wholesale distribution, the growth of imports, and greater competition among urban retailers.[13]

In view of their comparatively low wages, the hours of labour required from farmworkers were excessively long, and came to be regarded as anomalous as hours in factories, mines and workshops shrank characteristically to about 54–6 a week by the 1880s.[14] Farmers in the area of Wigtown

(Cumberland) thought their men unsettled by meeting those employed in mines and works where both higher wages and shorter hours could be had; and in Anglesey and in the Pwllheli district during the 1880s contact with weekend sojourners from the Caernarvonshire slate quarries excited the farmworkers to press successfully for a modest reduction in their hours.[15] But in general they remained remarkably long, and everywhere stockmen were called upon to put in hours which, averaged over a year, were higher even than those of the labourers. Sunday work, in particular, was disliked since usually no compensation was given, and as a consequence it was noticed that, in spite of their higher wages, cowmen were sometimes harder to obtain than ordinary labourers. Resentment at what a Lincolnshire worker called 'constant grind, month in and month out, with never an hour to call their own' was common, but unlikely to be met with much sympathy by employers. From one midland district in 1913 it was reported that farmers voted that the District Council should keep their roadmen at work until 4 p.m. on Saturdays (urban roadmen left at 1 p.m.) 'or else the farm labourer will want to leave also early'.[16]

Much contemporary comment focused on the deficiencies of cottage accommodation. Many horrific details were revealed in the sanitary inquiries of the early and mid-Victorian period, and although Canon Girdlestone, for one, discerned some improvement by the 1880s, inadequate provision remained in 1913 'a potent cause of migration to the towns'.[17] Until the Union Chargeability Act (1865), individual parishes bore the responsibility for their own settled paupers so giving a clear incentive to landlords to strive to reduce the number of cottages wherever possible. But although the Act destroyed the rationale for this, in no way did it encourage landlords to improve cottages or build additional ones. The advent of Sanitary Boards in the rural districts after 1875, and their successors, the Rural District Councils (from 1894), had little impact on the problem, which in the last analysis had economic rather than institutional roots. In a word, cottages were scarce because it was unprofitable to build them for a class of occupiers whose ability to pay rent was so limited. It was agreed that, in general, cottages belonging to major landowners were decidedly superior, but the return on capital expended was so low that the cottage building or improvement was a quasi-philanthropic activity of a kind in which lesser proprietors or the very numerous class of petty 'house-dealers' were unlikely to engage. In 1913 it was estimated that there was an absolute shortfall corresponding to some 6.5 per cent of the existing stock, and that if cottages unfit for human habitation were condemned, a further 5 per cent would be required.[18]

From the labourer's standpoint there was more to the cottage question than amenities or even rents, which by urban standards were very low. Although there is no record of cottages in model villages ever being difficult to let, and indeed testimony comes from Puddletown (Dorset) that news of a vacancy could excite village gossip into convulsions for weeks, such accommodation did not meet with universal approbation. As she passed through one such village, 'Lord so-and-so's place . . . with three bedrooms to every house and a pump to supply water to each group of cottages', Flora Thompson was told by her father that only good people were allowed to live there; a situation corresponding, perhaps, to that on Lord Wantage's Berkshire estate where, if there was no sign of squalor and idleness, 'they daren't blow their noses . . . without the bailiff's leave', in the view of one jaundiced observer.[19] Except where the labourer dealt directly with the landlord or his representative, cottages were predominantly sub-let through farmers and invariably tied to the holder's job. In 23 per cent of the parishes examined by the Land Enquiry Committee, the ordinary labourer received a 'free' cottage as part of his earnings, and where men were in charge of animals, the figure rose to 36 per cent. Not only was this seen as a species of truck, and a factor depressing wages; it implied a lack of independence. A Yorkshire labourer complained: 'There is a feeling about tied houses that it is not your house . . . you know that if the least little thing goes wrong, out you go at a month's notice or less.' The system also curtailed to some extent the freedom of the wife and children who (an Essex witness remarked) felt obliged to assist with housework and occasional work in the fields.[20] By contrast the town labourer, however poor, could usually call his home his own as long as he paid his rent.

The spread of elementary education exerted an influence believed by many employers to be pernicious. In Devon, for example, sixteen schools had been founded, chiefly under the auspices of the National Society, during the 1820s, and by 1870 the total number of voluntary schools in the county had reached 365.[21] Such a trend was general in England to a greater or lesser extent, although it was slower to take effect in Wales, and was matched by a prodigious growth of Sunday Schools. Yet scepticism about the value of education above the barest minimum remained well entrenched among farmers and landlords, and in some respects stiffened as facilities improved. Whether or not the school authorities took 'too scholarly a view of education', and whether the 'steady grind of bookwork' was calculated to give children a distaste for country life and divorce them from their surroundings, as Collings contended, was a moot point; it was certainly an argument capable of being exaggerated, as Richard Heath reflected on encountering youths from Shottery and from Hampton Lucy who did not admit to ever having heard of William Shakespeare.[22] The curriculum apart, a characteristic view was that a lad became accustomed to a warm room and dry feet; 'when he comes out he does not like a cold north-easter with sleet and rain, and mud over his boot-tops'. At Bunnington in the Vale of Wrington parents were drawn to send their children chiefly by the prospect of being able to obtain, by the rector's recommendation, good situations for the more promising of their children, 'sometimes as gentlemen's servants . . . pupil teachers, or clerks in merchants offices [or] porters'. All too often illiterate parents were so proud of their offspring that their sole aim was to 'put them to some trade or occupation other than farming', and from Lincolnshire a schoolmaster reported that he had often heard boys say, 'I'll never be a farmer's drudge if I can help it.'[23]

This comment takes us to the heart of the matter. What was so often decisive in influencing men to move out of agriculture into alternative employment or to the towns was what Acland summed up as a 'want of outlook', or in the words of the *Report on the Decline of the Agricultural Population* (1906) 'any reasonable prospect of advancement in life'. An intelligent labourer might become a master of his craft by the age of 21, 'but after rising to the position of horse-keeper or shepherd, or perhaps, foreman, there is very little further outlook'.[24] Precisely the same point had been made by Sir George Nicholls during the 1840s,[25] and whilst it was no doubt true that 'in no other industry does the workman require more skill, and in few . . . is such a variety of skills needed',[26] this was very evidently insufficiently recognized and rewarded. Thus about Glendale (Northumberland) a mere railway porter would 'consider himself socially superior to a hind, and . . . a hind's daughter would consider that she was bettering her position by marrying anyone not connected with agricultural employment'.[27]

## The drift to the towns

That the movement out of agriculture and migration to towns were related but by no means synonymous processes is immediately obvious from the fact that the workforce on the land fell substantially whilst the population of predominantly rural areas actually rose to some extent. Quarrying and brickmaking were common sources of alternative employment, notably in the Peterborough district, whilst in the neighbourhood of Southwell (Nottinghamshire) many young men were tempted to the nearby pit-banks; and in South Wales the remarkable development of the Glamorgan coalfield had caused farmers to become by the 1880s dependent on a regularly renewed influx of labourers from adjacent English counties.[28] Here and there substantial railway workshops presented another alternative. At Crewe, where the works employed 6,800 by 1891, the manager estimated that three-quarters were men who 'would otherwise have been available for neighbouring land', whilst at Ashford (Kent), where 2,000 men worked in the 1880s, the chairman of the Metropolitan Railway Company remarked that their apprentices were as a rule sons of agricultural labourers, adding that 'the labourers themselves come and work and perhaps their sons rise to be mechanics and fitters and moulders'.[29] Leaving aside establishments on this scale, employment at railway stations and in the rural police forces was frequently remarked on. Such changes of employment would necessitate giving up a tied cottage, but they need not entail leaving the village or at any rate going beyond the nearest small town. Similarly the availability of dockyard work at Pembroke and Milford Haven sustained moderate population growth even as numbers in the surrounding rural parishes declined.[30]

The dangers of exclusive emphasis on the condition of the agricultural labourer, and of assuming too close a correspondence between the shift out of agriculture and the drift to the towns, are made apparent again when we bear in mind that a majority of urban immigrants were females. Ravenstein recognized that at least over short distances women were more migratory than men, and it should be recalled that as late as 1911 domestic service remained by far the largest single occupational category. This was reflected in a differential sex ratio, especially in the 15–19 age group which then stood at 864 and 1,044 females per 1,000 males in rural and urban districts respectively.[31] Moreover, a large part of the efflux was made up of rural craftsmen. Saville has examined the case of Rutland, although it should be borne in mind that few if any other counties were so overwhelmingly rural in character, and that variations in census procedures do not permit exactitude.

**Table 4.2** Numbers of rural craftsmen in Rutland, 1851 and 1911

| *Craftsmen* | *1851* | *1911* | *% change* |
|---|---|---|---|
| Millers | 63 | 22 | −70 |
| Brickmakers | 38 | 15 | −61 |
| Sawyers | 33 | 10 | −70 |
| Cabinet makers | 31 | 10 | −65 |
| Coopers and Turners | 15 | 2 | −87 |
| Wheelwrights | 74 | 42 | −43 |
| Blacksmiths | 116 | 83 | −28 |
| Building trades | 514 | 415 | −19 |
| Saddlers | 31 | 24 | −23 |
| Tailors | 173 | 63 | −64 |
| Shoemakers | 236 | 138 | −42 |
| Population of Rutland | 22,983 | 20,346 | −11 |

Source: Saville, 1957, 74.

The factors influencing non-agricultural elements in the rural population to join the drift to the towns have been much less discussed than those relating to agricultural labourers. But it seems clear that the emerging railway system was significant, not as is sometimes naively assumed because of its ability to move people from one place to another more efficiently, but on account of its role in fostering a national market. The competitive power of large-scale urban enterprise was powerfully reinforced and brought about the transfer of many rural crafts and small country industries to the towns.[32] Whilst endorsing this view, a more recent study has suggested that railways exercised an influence on men's imaginations; excursions, shopping expeditions, and similar diurnal mobility may well have made the prospect of living and working elsewhere less painful to contemplate.[33] This must have been more appreciable among non-farm employees, who would have found it much easier to take advantage of such facilities.

The processes involved in the flow from villages into small towns, and from small towns to larger ones, were thus more complex than is generally realized. For example, we have little idea of how frequently movement occurred by stages, and how this related to the declining numbers in agriculture. In the case of mid nineteenth-century Preston, Anderson discerned a considerable amount of two-step migration in the sense that villages contributing to the growth of the central nucleus in turn drew on places further afield, after the manner described long ago by Ravenstein and by Redford. But there were only slight traces of deliberate two-step migration by the same individuals, except among those born at a distance, who made up only a small proportion of all immigrants.[34] How typical this was is not at present known, and illustrates the kind of behavioural question which existing studies based on aggregative census materials and the analysis of net flows can hardly begin to approach. It is clear that the farm labourers and rural craftsmen came under different pressures. For the former they might be met simply by a change of occupation with or without a move to the nearest town. This would perhaps satisfy many men of limited outlook and ambitions, but some would go further; from Buntingford (Cambridgeshire) it was reported that 'all the quick-witted ones go to London'.[35] Likewise among craftsmen in larger villages and small towns, some would remain and others envisage advancement from removing after completing their apprenticeships. As Llewellyn Smith remarked, 'There are villages and country towns which may be described as breeding grounds for journeymen for the great cities.'[36]

Economic factors apart, we should not overlook the appeal of what the same author described as 'the contagion of numbers, the sense of something going on'. Thus, carters and busmen were apt to say that the busy streets of London were simply more interesting than country roads, and it would be unwise to overlook the temptations of 'the theatres and the music halls, the brightly-lighted streets and busy crowds: all, in short, that makes the difference between Mile End Fair on a Saturday night, and a dark and muddy country lane, with no glimmer of gas and nothing to do'.[37] Then there was the attraction of the girls, who in a sense blazed the trail. To say that at Lark Rise in the 1880s there was no girl over 12 or 13 living permanently at home may exaggerate but does not mislead.

For the vast concentrations of female domestic servants in towns unquestionably served to 'act as a magnet to the lads they leave behind them'.[38]

## Implications of the urban influx

Most of our systematic evidence on the occupations taken up by migratory countrymen relates to the great cities, particularly London, and was collected in response to the re-emergence of the 'condition of England' question during the 1880s and after. In what was by his own admission a 'scattered notice' of the London trades, Llewellyn Smith concluded that in general immigrants from the country were under-represented on the London docks, where labour was recruited from the native-born or those resident for some considerable time. On the other hand, the building trades were 'most overrun by countrymen', whilst they were also employed in large numbers on the railways and abounded on the roads in such roles as carriers or omnibus drivers.[39] More systematic information drawn from the 1881 census by Stedman Jones for the London labour force as a whole shows that while 50 per cent was the general average born in the provinces, this was exceeded among the police (87 per cent), gardeners and railway labour (78), railway service (69), brewery workers (61), carpenters and domestic servants (59), busmen (58), gas works servants (57), builders (56), masons (54), and municipal labourers (52), etc. On the other hand, the most highly skilled London trades (such as bookbinding, printing, jewellery, manufacture of musical instruments) were dominated by native-born Londoners, along with occupations like dock labour and coal-heaving.[40]

It should be noted that from the metropolitan standpoint, all immigrants (including, let us say, a mason born and trained in Birmingham) were 'countrymen'. No doubt many from provincial towns great and small, and from large villages, came to offer the skills and services in which they already had experience. The former agricultural labourer necessarily undertook an occupational change which, given his starting-point, could hardly fail to be a species of promotion. Llewellyn Smith compared the former and present occupation of some 500 London immigrants born in a group of villages and small market towns in Hertfordshire and south Cambridgeshire, showing that the proportion describing themselves as labourers fell from 640 to 169 per thousand, while the largest gains were made under the headings of military service (+108), railway employment (+87), menservants and grooms (+59), carmen (+49), gardeners (+35), domestic service (+41), police (+34), porters (+22), and soap, gas, and chemical works employment (+21).[41] Likewise Wilson Fox, who assembled information on 19,000 regularly employed countrymen in London and 27,000 in the great provincial cities in 1906, found that former agricultural workers accounted for 23, 21, 35, and 37 per cent of the labour force of four London breweries; 25 per cent of the employees of the South Metropolitan Gas Company; 18 per cent of goods porters at the Great Northern Railway Terminus and no fewer than 47 per cent of their stablemen; and 22 per cent of the work people employed by the sixteen largest English Municipal Corporations.[42]

With a large proportion of immigrants undertaking similar work at better wages and others undergoing elevation in the manner described, there is no reason to expect that, even discounting individual success stories, country immigrants would feature disproportionally in the statistics of distress. Indeed, figures brought forward by Llewellyn Smith in the 1880s and by Wilson Fox in 1906 suggested that the overwhelming bulk of those in indigent circumstances were town-born, both in London and the great provincial towns.[43] Stedman Jones has drawn attention to the fragility of such statistics, and emphasized that the oft-repeated statements by employers of a preference for the countryman were based as much on his presumed docility (London men were described as 'shuffling, lazy and know too much' by one brewer) as on his greater vitality and productivity. No doubt, as he suggests, the 'theory of urban degeneration' was an inadequate, incomplete, or inappropriate explanation of the weak competitive situation of the London-born, and it seems likely that the superiority of provincials over the debilitated armies of 'pure-bred' Londoners was exaggerated.[44] But at least there is no evidence for the contrary view – that immigrants were especially disadvantaged in the urban labour market, or failed to secure a due share of what work was available, having regard to their background and experience.

Were there not, however, other social costs attendant upon migration? Obviously the inflow must have put pressure on the urban infrastructure and especially on the housing stock, since without it development could have been at a more leisurely pace and the quality of the urban environment might have been higher. Whether it would have been is impossible to say. Moreover, the influx tended to call into being its own supply of housing, especially as real incomes rose more rapidly in the later years of the century. Leaving aside the case of the Irish who patently were left with the dregs of urban housing, it is significant that in London in 1881 a greater proportion of provincials were to be found in the rapidly expanding suburbs than in the central districts such as Bethnal Green, where little more than 12 per cent had been born outside London. All the evidence suggests that the inner metropolitan slums were 'settlement tanks for submerged Londoners' rather than reception centres for provincial immigrants.[45] Even at Lincoln it was apparent by the Edwardian period that most of the working class lived in new housing areas rather than in the older parts of the city.[46] This is consonant with the statistics of distress.

However, it was frequently suggested that migration to the great towns was destructive of personality, family ties, and wholesome values. A theme popular in Victorian melodrama was London's 'homeless poor – their crime, drunkenness and nostalgia for the lost life of the village', a scenario which no doubt reflected some actual case-histories but hardly the fate of the majority.[47] Only one element in this stereotype is amenable to examination, namely the supposition that migration was destructive of family ties. Whilst we must allow, in the words of a *Morning Chronicle* investigator, 'for a great number of shades of family disruption' and presume that in the long term immigrants would cease to perform important functions within the ongoing family system in their birthplaces, recent research shows that at any rate in the first generation contacts with kin were valued and actively pursued.[48] A new arrival needed a roof over his head, a job, and someone to help him to adjust to the new environment, and kin, along with co-villagers, were his main recourse. Among 202 cases traced by Llewellyn Smith, about one half had definitely secured a place in town before leaving the country, and, he remarked, once a country nucleus was established in London, it grew 'in geometric ratio by the importation of friends and relations'.[49] There is also evidence of reciprocal services in the form of numerous examples of migrants seeking assistance from their country relations and sometimes returning there in times of crisis or distress.[50] Not a few, like 'John Jarman' of 'Little Guilden', [51] retired to their villages, so that the flow was by no means uni-directional. Indeed, in every respect including the transfer of people, town and countryside were highly interdependent.

## Overseas migration

Table 4.3 shows that with the exception of the 1840s, when Irish immigration was in full spate, net losses from the residual rural areas as defined by Cairncross exceeded the net gains by urban districts in every decade from 1851–61 to 1891–1901, and especially in the 1880s. It serves to draw attention to the role of overseas emigration as an alternative to

**Table 4.3** Urban net gains and rural net losses 1841–1901

| | *1841–51* | *1851–61* | *1861–71* | *1871–81* | *1881–91* | *1891–1901* |
|---|---|---|---|---|---|---|
| A Rural losses ('000) | 443 | 743 | 683 | 837 | 845 | 660 |
| B Urban gains ('000) | 742 | 620 | 623 | 689 | 238 | 606 |
| C Imputed net loss by overseas emigration from rural areas (A–B) | 123 | 60 | 148 | 607 | 54 | |
| D C as a % of A | | 17 | 9 | 18 | 72 | 8 |

Source: Cairncross, 1953, 70.
Notes: (i) The rural losses are equivalent to the sum of the 'North' and 'South' figures in Table 4.1 above. (ii) In 1901–11 both the rural and the urban areas showed net losses by migration, i.e. 295,000 and 207,000 respectively. (iii) Colliery districts are included with the urban areas.

permanent settlement in the towns, although the matter is vastly more complex than might appear at first sight.

Even before the 1834 Poor Law Amendment Act magistrates were looking with favour on emigration as a remedy for the 'surplus population' problem of the southern counties, for example in Biddenden, Cranbrook, and other parishes of Wealden Kent.[52] The authorities charged with administering the New Poor Law espoused this policy, and between 1836 and 1846 assisted some 14,000 persons to emigrate to the colonies, chiefly Canada, from England and Wales,[53] yet the policy was never prosecuted with the vigour that might have been expected. As early as 1837 it had occurred to many guardians in East Anglia that their interests as ratepayers and employers were in conflict,[54] and after 1850 the role of the Poor Law authorities sharply diminished. Another official body was the Colonial Commissioners of Land and Emigration. This was set up in 1842 to aid the emigration of selected persons using funds raised from the sale of Crown lands in the colonies, and by 1869 it had assisted well over 300,000 United Kingdom citizens (including Irish) to emigrate, chiefly to Australia.[55] However, while such sources of assistance were probably crucial in the case of agricultural labourers, it should be born in mind that, overall, the bulk of British emigration was privately financed and often independent. It also flowed mainly to America, which during the first thirty years of Victoria's reign absorbed some 3.5 million persons from the United Kingdom, against 0.75 million by Canada and 1 million by Australia and New Zealand.[56]

Table 4.3 suggests an apparently much greater rate of loss by overseas migration from rural areas in the 1870s, coinciding with a recorded 140 per cent increase in the number of adult farm labourers, shepherds, gardeners, and carters, when the British emigration statistics of 1861–70 and 1871–80 are compared.[57] These years saw the emergence of the first successful agricultural trade unions, which, following precedents set by others, took vigorous steps to promote emigration on the presumption that the bargaining power of those who remained would be enhanced. Thus January 1874 saw the departure of 410 Kentish emigrants at the instigation of the Kent and Sussex Labourers' union, and in 1879, against the background of a lockout, a farewell tea at the Skating Rink in Maidstone was held for a similar number, shortly to depart for New Zealand in a chartered Dutch vessel, the *Stad Haarlem*. 'They wanted to "get on", these gentle civil-spoken southern agriculturalists' commented a *Daily News* reporter. 'There is no blatant mob-oratory ring about their modest aspiration. "We should like to see our children better off than we have been." '[58] Yet the encouragement and assistance of the unions did not endure for more than a few years. Although the funds of the National Agricultural Labourers' Union benefited by 3s. per adult from the Canadian Agent General (a figure more than offset by disbursements in the form of assisted passages), the activities of some of its officials were not above suspicion. Some, such as the chairman of the Oxford District of the NALU, were in receipt of commissions and retainers, and on a train journey to London Joseph Arch, who had an ambivalent attitude to the matter, was outraged to overhear the comment, 'He's selling them like cattle to the foreigners.'[59] After 1881 there were clear signs of waning interest. Support for the unions was falling and they felt increasingly less able to assist.

Yet, as the most recent research has confirmed, the efflux was still more impressive in the 1880s. Overseas emigration from thirteen among thirty-four rural counties definitely peaked during this decade, notably in Lincolnshire, Norfolk, Suffolk, rural Wales, and the border counties. Moreover, the number of emigrants whose occupations were recorded as agricultural workers continued to rise; at one-sixth (a minimum estimate), they were now over-represented in proportion to their share of the total occupied population. However, over twice as many males and four times as many females who left the rural counties in these years were internal migrants, many of them destined, no doubt, to fill the places of townsfolk born and bred, who, even at this juncture, contributed much more to the outflow.[60] In the ensuing decade, the 1890s, there was a marked decrease in total emigration and of overseas emigrants born in the rural counties although, because the number who were natives of urban counties fell by more, the proportion who were rural migrants rose marginally; at the same time there was a small decline in the rate of rural-urban migration, which limits the extent to which overseas and internal migration can be viewed as simple alternatives for the rural population.[61] After 1900 there was a renewed boom in overseas migration, directed particularly towards Canada. We know that the outflow did not leave rural areas unaffected: men were reported to be leaving by the hundred from around Dorchester and Canterbury just before the War and a Board of Agriculture report of 1913 instanced the case of Snaith in the East Riding, where the church choir was said to suffer from continual attrition and the cricket club had lost twelve members by emigration since 1912.[62] Even so, it is certain that the bulk of Edwardian emigrants were drawn from the towns, and it is significant that net losses from urban as well as from rural areas were in evidence for the first time.[63]

## Remedies

The first signs of concern about the loss of rural population came in the form of sporadic complaints from landlords and farmers about seasonal labour shortages, chiefly from the 1850s onwards.[64] Towards the end of Victoria's reign the rural exodus began to engender more comment, and in the aftermath of the Boer War it featured regularly in discussions of 'national efficiency', with authors such as Arnold White deploring the implications of sturdy countrymen giving place to 'white-faced workmen living in courts and alleys'.[65]

There was, it seemed, no answer to be found. In so far as wages in agriculture were the key, we can discern a definite tendency for the real wages of full-time employees to advance in the later nineteenth century, and the regional wage disparities so evident in Caird's day diminished without disappearing altogether.[66] Yet by any urban or industrial yardstick the labourer remained poorly remunerated. In evidence to the Land Enquiry Committee the secretary of the Agricultural Labourers' Union lamented that 'forty years of experience has convinced me that the labourers cannot get a living wage by Trade Union effort alone', and it is significant that the report of the committee advocated the kind of minimum wage legislation which in fact was first introduced for farm labourers in 1917.[67]

The cottage question continued to generate fitful discussion without much sign of progress, and indeed the last pre-war issue of the *Journal of the Royal Agricultural Society* included yet another essay on the subject.[68] Suggestions that education in rural schools should be reorientated towards subjects believed likely to contribute to the stability of country life were aired by figures such as the eminent scientist and Liberal MP Sir Henry Roscoe, and by school inspectors anxious to develop the 'seemingly instinctive' interest of country children in 'the green earth and its feathered and four-footed tenantry', but to little avail.[69]

The provision of allotments as a means of easing the situation of the labourer had a long history, and, defined as holdings capable of being cultivated by the labourer without help other than from his family and in his spare time, they were already to be found in some 42 per cent of English parishes as early as 1833.[70] Allotments continued to be recommended in various official reports and by sundry publicists throughout the century, yet it would seem were far from being the obvious answer that some of their more enthusiastic advocates supposed. For, if remote from the village, on poor land, or subject to restrictions in use, they were unlikely to meet with the labourers' approval. Thus in Kent the system of 'letting land to the poor at double or treble its value' was reckoned likely 'to exasperate the lower orders still more against their superiors'; whilst forty years on F. G. Heath remarked that around Bridgwater the rents charged by farmers were more than four times those paid to their landlords – a situation, he added, general in the west of England.[71] Although in 1882 the Allotments Extension Act laid down that the trustees of charity lands in any parish should be directed to provide allotments, the issue did not assume any political importance until 1885, when large numbers of rural constituencies were won by the Liberals (including Joseph Arch in North-West Norfolk) on an agrarian programme which included allotments under the slogan 'Three acres and a cow'. Further legislation in 1887 and after empowered local councils to acquire land compulsorily for the purpose, but they were in many cases very slow to move, being dominated by farmers and landlords hostile to the very idea. Moreover, there is reason to think that among the labourers the desire for allotments was 'very variable'.[72]

Obviously the allotment issue shaded over into the smallholding question and both movements could be said to have originated in a sense from the impolitic social consequences of enclosure. But the aims of the smallholdings movement were quite different, looking to the re-establishment of a class of peasant proprietors wholly occupied on their own holdings. Advocates of such a step, who were especially vociferous in the 1880s, were not infrequently moved by anti-landlord prejudice, by a distaste for the 'monopoly' of land evidenced in the New Domesday Survey of 1870, and by a conviction that continental systems of land tenure were by no means as inefficient as had been assumed by the protagonists of large-scale capitalist agriculture in Britain. Labour leaders thought that such a programme might stem the flow of labour into towns and so mitigate any threat to urban wage levels, whilst sentimentalists saw it as a step towards the re-creation of Merrie England.[73] More significantly, Tory politicians such as Henry Chaplin and Lord Salisbury came to feel that a peasant proprietary could constitute a bulwark against revolutionary change, their ideas in this respect corresponding with those of the chief author of the movement, Jesse Collings, who freely conceded that his proposals were 'in the true and not in a party sense of the word conservative in the highest degree'.[74]

It was no doubt this coincidence of interest that made possible an Act of 1892 empowering county councils to create smallholdings for purchase by instalments; and that of 1908 which enabled them to provide holdings for sale or for letting. Whatever the merits of peasant proprietorship or of the smallholding ideal (and they did not go uncontested since it was possible to raise some very powerful counter-arguments) comparatively little was accomplished in practice, in part due to the reluctance of county councils actually to use their powers. The general position with respect to the labourer's access to land was reviewed by the unofficial Land Enquiry Committee of 1913. It was estimated that about two-thirds of all villages in England had allotments – a proportion, it should be noted, considerably higher than in the 1830s; whilst since the 1908 Act some 10,000 smallholdings had been provided by county councils, corresponding to 3.3 per cent of all holdings in the range 1–50 acres. There was, the committee maintained, a 'large unsatisfied demand' for both.[75]

Such were the remedies proposed to stabilize rural life and the rural population. It is unlikely that they had any significant impact on the drift from the countryside. Yet in the twenty years before the holocaust of the

First World War the efflux showed definite signs of abatement (see table 4.1) and the number of persons described as agricultural labourers actually rose slightly between 1901 and 1911. It is not particularly easy to account for this. One explanation, advanced by Cairncross, refers to falling rates of national increase in the rural areas (the consequence of age-selective migration) which necessarily reduced the size of the pool of potential movers.[76] But it has been shown more recently that there was no fall in the number and proportion of young adults (15–34) in most rural counties down to the 1890s, and it is well known that agricultural labourers' fertility, as gauged in 1911, remained high and was exceeded only by coal-miners.[77] No doubt, too, aggregate rural population changes were increasingly affected by suburbanization in some districts. Issues such as these are crying out for more attention than they have so far received, at a more detailed level of analysis than that of the county. In the meantime, the other explanations advanced by Cairncross to account for the evident diminution in the rate of loss are plausible and include the following points: first, a degree of revival in British agriculture after 1900 as prices improved. Second, the operation of the law of diminishing returns 'to protect what was left of rural industry just as it was operating to protect agriculture, and third, a pause, albeit temporary, in changes in agrarian structure because the limits of known technique had been reached.[78]

## Notes

1 See chapter 2.
2 Thus, where $P_1$ is the first census population and $P_2$ that of ten years later, B is the number of births and D the number of deaths occurring in the intervening period, the observed inter-censal increase is $P_2-P_1$, natural increase is $B-D$, and the net migration component is $(P_2-P_1)-(B-D)$. Baines, 1985, ch. 4., offers a full discussion of sources and techniques for the study of inter-county migration flows.
3 See, for example, Ogle, 1889, and the critical comments in Saville, 1957, 64–5.
4 Cairncross, 1953, 78. In the main his arrangement of the published source material is relied upon here, but his conclusions do not differ in any important respect from those arrived at in subsequent studies, in particular Saville, 1957, and Lawton, 1973, although both offer more detail.
5 Cairncross, 1953, 74–5; Lawton, 1973, 211.
6 Hasbach, 1966, 256, 258.
7 F. G. Heath, 1874, 250.
8 Kerr, 1968, 238–42; Ashby and King, 1893, 6.
9 BPP 1893–4 XXXV [C. 6894-V],92; [C.6894-VI], 8; BPP 1893–4 XXXVII [C. 6894-XXV], 40.
10 See chapter 2.
11 BPP 1893–4, XXXV [C. 6894-II], 45; and see Horn, 1976, 76.
12 Bellerby, 1953, quoted in Saville, 1957, 13.
13 Hunt, 1973, 88–100.
14 Bienefeld, 1972, 82, 122.

15 BPP 1893–4 XXV [C. 6894-III], 144; BPP 1893–4 XXXVI [C. 6894-XIV], 130, 147.
16 Land Enquiry Committee, 1913, I, 16–17.
17 Ibid, 80; BPP 1884–5 XXX [C. 4402-I], 635.
18 Land Enquiry Committee, 1913, I, 132–3.
19 F. G. Heath, 1874, 36–7; Flora Thompson, 1939, 335–6; Havinden, 1966, 115.
20 Land Enquiry Committee, 1913, I, 30, 138–9, 144.
21 Sellman, 1967, 25.
22 Collings, 1908, 23–5; R. Heath, 1893, 235.
23 BPP 1906 XCVI [Cd. 3273], 32, 34; F. G. Heath, 1874, 135; Thirsk, 1957, 323.
24 Land Enquiry Committee, 1913, I, xxxiv; BPP 1906 XCVI (Cd. 3273], 11–21.
25 Nicholls, 1846, 3–4.
26 Collings, 1908, 382.
27 BPP 1893–4 XXXV [C. 6894-III], 103.
28 BPP 1906 XCVI [Cd. 3273], 31; BPP 1893–4 XXXV [C. 6894-I], 114; BPP 1893–4 XXXVI [C. 6894-XIV], 8.
29 BPP 1893–4 XXXV [C. 6894-IV], 109; BPP 1884–5 XXX [C. 4402-I], 449.
30 Gilpin, 1960, 3–7.
31 Ravenstein, 1885, 198–9; Saville, 1957, 110.
32 Cairncross, 1953, 75.
33 Hunt, 1973, 266–71.
34 Anderson, 1971b, 21, 25–6; and see Ravenstein, 1885, and Redford, 1926.
35 BPP 1893–4 XXXV [C. 6894-II], 149.
36 Llewellyn Smith, 1904, 141.
37 ibid, 75.
38 Flora Thompson, 1939, 163; BPP 1893–4 XXXV [C. 6894–1], 18.
39 Llewellyn Smith, 1904, 81–2, 90, 92, 96, 98, 141.
40 Stedman Jones, 1971, 137.
41 Llewellyn Smith, 1904, 140.
42 BPP 1910 XLIX [Cd. 5068] Appendix J, 729.
43 ibid, 724–7; Llewellyn Smith, 1904, 84–5.
44 Stedman Jones, 1971, 132, 134–5, 143–4.
45 Llewellyn Smith, 1904, 66, 121–2; Dyos and Reader, 1972, 372–3.
46 Hill, 1974, 301–2.
47 See, for example, Dyos and Wolff, 1972, ch. 8.
48 Anderson, 1971b, 66, 152–8.
49 Llewellyn Smith, 1904, 133–4.
50 Anderson, 1971b, 158.
51 Llewellyn Smith, 1904, 131–4.
52 Melling, 1964, 175–9; Johnston, 1972, 99–101.
53 Erickson, 1976, 127.
54 Digby, 1975, 80–2.
55 Erickson, 1976, 122.
56 Woodruff, 1934, 363.
57 Horn, 1972, 100.
58 Arnold, 1974, 87; and see p. 83 of the same journal.
59 Horn, 1972, 94–5; Arch, 1898, 83.
60 Baines, 1985, 76, 77, 192, 203, 204, 205, 238 (table 8.9), 264, 281.
61 ibid., 183, 184, 187, 192, 238, 239, 281.

62 Aronson, 1914, 19; Board of Agriculture and Fisheries, 1913, 10.
63 Cairncross, 1953, 71.
64 E. L. Jones, 1964, 328–9.
65 White, 1901, 96; and see Stedman Jones, 1971, ch. 6.
66 See chapter 2.
67 Land Enquiry Committee, 1913, I, 42, 67.
68 Allen, 1914.
69 Ausubel, 1960, 200; Collings, 1908, 25–30; and see Land Enquiry Committee, 1913, I, 436–44.
70 Barnett, 1967, 163.
71 *The Labourer's Friend*, 1835, 213; F. G. Heath, 1874, 75–6.
72 BPP 1893–4 XXXV [C.6894-VI], 29.
73 Orwin and Whetham, 1964, 332.
74 Collings, 1908, xxiii.
75 Land Enquiry Committee, 1913, I, 188–9, 191, 229.
76 Cairncross, 1953, 75.
77 Baines, 1985, 185, 206; Innes, 1938, 42.
78 Cairncross, 1953, 75–7.

# 5

# Rural culture

Charles Phythian-Adams

The year of the Great Exhibition, when Britain celebrated its primacy as the most advanced technological nation in the world, also marked the numerical supremacy of town populations over those of the countryside. Yet even in the industrialized Leicester of 1851, there lived a 'wise woman', who was being consulted for her magical prowess from as far afield as Rutland.[1] Nor was she unique. 'Wise men' or 'cunning men' were still at work in or near other urban centres: Black Jock of Newcastle; Oakley of Tunbridge Wells in the 1860s; Snow and Tuckett at Exeter during the 1880s.[2] One of the more celebrated was 'Au'd' Wrightson of Stokesley in Yorkshire (*fl. c.* 1820) who had at least one successor; another was the wise man who lived between Hereford and Bromyard in the 1860s.[3] It is likely that market towns in many of the traditional areas of the country were the homes of such figures during at least the early decades of Victoria's reign.

These people were not simply the primitive vets of their age. Their powers were widely seen to be semi-magical: they charmed away the ills of both ailing farm animals and human beings and used spells and rituals in so doing. Above all, their aid might still be invoked when misfortunes were otherwise so inexplicable in normal terms that bewitching was suspected.

For the fear of witchcraft was not yet dead. Cecil Torr's father noted in 1844: 'Witchcraft a common belief to this day at Lustleigh (Devon), and prevalent even among the better informed classes.'[4] Writing from his own long and deep knowledge of the north York moors, the Rev. J. C. Atkinson stated categorically in 1891:

> Fifty years ago the whole atmosphere of the folklore firmament in this district was so surcharged with the being and the works of the witch, that one seemed able to trace her presence and her activity in almost every nook and corner of the neighbourhood.[5]

In 1871, the affliction of the so-called woman-frog of Presteigne, who only ventured out of the house to hop to and from the Primitive Methodist chapel, was solemnly explained to Francis Kilvert as the outcome of a curse laid on the wretched woman's mother when she was pregnant.[6] As late as the 1880s, a Hastings man was recommended to singe the wrists of his bewitched wife with a poker before the hearth 'to make the evil spirit fly up the chimney'.[7] Writing about the same decade, Miss M. A. Courtney claimed that 'belief in witchcraft in West Cornwall is much more general than most people imagine'.[8] Long after witches were publicly exposed and punished, the fear of them lived on in conservative areas. Even in 1939 Christian Hole

> was told of several witches who live in a village near Ilminster; one woman there takes the precaution of placing crossed knives on her stairs whenever the moon is full to defeat the spells of one of them whom she particularly fears.[9]

If, as a result, a whole range of protective devices against evil – from horse-shoes to rowan twigs – was still widely employed in some regions during part of Victoria's reign, no less bewildering to the modern mind are many of the popular cures which similarly continued to be used. From the Lake District come newspaper reports in 1866 and 1876 (and from Cornwall in 1885) of ritual remedies against contagious abortion or brucellosis amongst calving cows, which involved either the burial alive or roasting alive of young calves.[10] Elsewhere popular cures for whooping cough still included, for example, at best taking the sufferer into the mouth of a cave or through the arch of a re-rooted bramble; at worst, either the consumption of a roasted mouse by the afflicted child or the placing of the head of a live trout or frog inside the child's mouth.[11] As late as 1902–3, a Devon farmer claimed that his malformed baby was as likely to be cured by being passed three times through a split ash tree at sunrise as by 'sloppin' water over'n in church', and added significantly that 'all folk do it'.[12]

The survival, well into Victoria's reign, of what has been usefully dubbed 'the prior culture' is thus hardly in doubt.[13] Its ritual expression was also more widespread than is sometimes supposed. A remarkable survey of what are often regarded as mummers' plays still extant in the nineteenth century, for example, associates them with roughly 800 places in England alone.[14] Their distribution indicates that such observances were not necessarily immediately vulnerable to the progress of either commercial farming or urbanization. These plays were acted far beyond the confines of Hardy's Wessex and were often performed by clusters of villages in close proximity to cathedral towns like Lichfield or even near an industrialized centre like Leeds. Above all it looks as though seasonal drama of this kind was a surviving cultural feature of the nucleated rural community in whatever region it might be found.

If vernacular culture was not yet moribund, neither was it monolithic. Different types of play, for instance, might be regionally restricted: to Lincolnshire and Nottinghamshire in one case; or to the North Riding, Durham, and Northumberland in another. There were even significant calendrical differences with regard to the performance of the same type of play. To judge from those places where the timing is established, the so-called Hero-Combat type was usually performed in Cheshire around All Souls, but at Easter north of the county boundary – in Lancashire, an adjacent part of Yorkshire, Furness, and southern Westmorland. In other parts of the country, by contrast, it was played between Christmas and Plough Monday. Still other areas – Wales on the one hand, and Bedfordshire, East Anglia, Essex, and Hertfordshire on the other – appear to have been largely innocent of any such activity.

A study of the distribution of an identifiable form of popular drama shows that it would be premature to assert that in the earlier nineteenth century traditional activities and beliefs were restricted only to the remoter parts of the highland zone (the extent and significance of which is apt to be depreciated by the average lowland-based historian). Work done on the subject as a whole is geographically so scattered that it is impossible to map the incidence of survival in our present state of knowledge. What is clear is that even counties like Suffolk and Sussex contained pockets of traditional culture down to the First World War at least.[15] *Jackson's Oxford Journal* of 1837 was able to claim: 'In no other part of the united kingdom, we believe, are these old English revels [i.e. Whit Ales] celebrated with such spirit, so much original character, as in the midland county of Oxford.'[16]

If the 'prior culture' was still widely entertained, so too were the springs of its inspiration. For all its faults, A. J. Ellis's great dialect survey, the rough outlines of which still seem to stand the test of time, indicated six major dialect divisions of the country in the 1870s (southern, western, midland, eastern, northern, and lowland Scottish), and no fewer than forty-two subdivisions or districts, each with 'a sensible similarity of pronunciation'.[17] Even within counties, linguistic diversity might be remarkable. In Somerset alone, no fewer than twenty-three – often very different – names are recorded for the scarlet pimpernel; and no fewer than thirty-three for the birdsfoot trefoil.[18] It can hardly be doubted that within the countryside problems of linguistic communication, together with short-range marriage patterns, still helped to conserve localized attitudes beyond the middle of the century.

Across the nation, indeed, local loyalties were expressed in such rhyming jibes as:

Tring, Wing and Ivinghoe
Three dirty villages all in a row,
And never without a rogue or two:
Would you know the reason why?
Leighton Buzzard is hard by.[19]

A recently discovered notebook recording life in Cornwall at the turn of the century claims that in that county there was probably a rude nickname for the residents of every parish, ranging from 'Morvah chitchats' through 'Madron squerts' to 'Sennen hoars'.[20]

In these circumstances it is hardly surprising that certain basic popular beliefs were also variously expressed in contrasted regions. The wood or foliage of specific trees, in particular, were held to be peculiarly efficacious in protecting against witchcraft, for example. In different parts of the country, therefore, primacy in this respect might be accorded to the oak or the ash, hazel or holly, rowan or hawthorn. All were associated with lightning in one way or another; all appear to have been used for ritual fires at Christmas or other times. Similar differences in expressing the same attitudes may be detected with regard to 'sacred' birds. In much of England and Scotland, for example, the robin and the wren were widely termed 'God Almighty's cock and hen', the latter being described as the 'Lady of Heaven's hen' over the border. In both areas some people still held that it was taboo to kill these birds or to destroy their eggs or nests. In Essex, Sussex, the Isle of Man, Wales, and (even today) southern Ireland, by contrast, the wren was deemed to be male not female, and was ritually hunted and killed around Christmas-time; where it was not, the outline of the custom often survived. The bird was still held to be sacred, however. It was considered particularly lucky to possess one of its feathers, and a frequently found song on such occasions began

The Wren, the Wren, the king of the birds
S. Stephen's day was killed in the furze.[21]

Ancient cultural traditions of this kind – and many more – thus helped to perpetuate those invisible barriers between one locality and another which together still contributed to the colour and diversity of the more custom-bound countryside in early Victorian England.

Even as late as 1837, therefore, it may be claimed that established conventions still dictated mental attitudes, however variously across the country. That said, a balance must be struck. It is an undeniable fact that many popular superstitions and observances had been whittled away already in the more agriculturally developed localities over the preceding centuries. In so far as there had ever been something approaching a framework of beliefs before the Reformation, much of this had been eroded subsequently by Protestant activity in its many forms; by state intervention in such a basic matter as the alteration of the calendar in 1752; by the impact of enclosure on communal farming rituals; by the anti-blood sports campaign; and perhaps even by the Napoleonic wars, which may have been a cultural watershed comparable to that of World War I a century later.[22]

Nevertheless, while certain observances had long been discontinued in many areas, and with them some or all of their attendant ritual detail, many of the mental attitudes on which such practices were founded do appear often to have survived. Only thus can we explain the fact that, despite the gloomy statements of the late Victorian folklorists (who anyway rarely interviewed their few informants systematically) on the imminent disappearance of their cherished subject-matter, a sensitive modern investigator like George Ewart Evans is able to elicit from old folk even today a truly remarkable amount of evidence for the survival of many old beliefs down to 1914.[23] Moreover, while public rituals clearly disappeared long before some of the superstitions and culinary customs related to them, the ideals of private ritual observance (particularly with respect to the defeat of witches or placating of fairies) were sturdily perpetuated in numerous folk-tales and songs. In these the traditional basic attitudes and beliefs continued to be interrelated in marked contrast to the confusing mass of separate superstitions collected by the early folklorists. It is worth noting that many of these tightly constructed tales have been recorded for the first time only in this century.[24]

If the essential medium of popular culture was oral, this in itself underlines the unchanging circumstances in which the old beliefs were perpetuated. Even by Victoria's reign the education of ordinary country folk had not progressed far. Such advances as had been achieved in medical and veterinary practice were far from always available to them. Above all, large numbers of small farmers, labourers, and rural industrial workers continued to live in primitive conditions on the very edge of nature and in close relation to their farm animals. For such people the vagaries of the climate, of the environment, and of disease in both human beings and animals were ever-present realities. In times of misfortune, especially, it is difficult to see how they could have fallen back on anything except what they took to be the conventional wisdom. Christian faith in itself would not cure the ailing family pig.

The major facets of this conventional wisdom, despite the local diversity of its expression, may be readily identified. Popular attitudes during much of the first half of the nineteenth century were still concerned with three matters. The first was the correct observance of established practices, whether these involved the customary prosecution of annual or day-to-day activities like planting or cleaning the house, the performance of private rituals, or obedience to certain taboos. Second, tradition taught that it was possible to anticipate the future in certain ways, and having done so either to take avoiding action where that was possible or, more often, to accept one's fate. Third, if dire and unanticipated misfortune did strike, however, even though the customary procedures had been observed, society could yet turn to the doings of the witch as the ultimate explanation. At root the mental attitudes involved were thus understandably obsessed with accounting for and meeting the omnipresent threats of destitution or dearth, disease and death.

For those who still clung to these beliefs (and it is impossible to measure precisely how strongly they held them by this period), all these preoccupations emanated from a world view that was essentially medieval in its conception. Its central assumption was the existence of supernatural forces in the human body, in nature and in the heavens. This said, it is necessary to emphasize that there is no convincing evidence known to the writer – either from this or the preceding three centuries – which might suggest that even conservative country people thought in terms of pagan deities or tutelary spirits. Phoebus, Luna, Ceres, and Pomona were figments of the classically educated imagination. Where not superimposed on the evidence by contemporary commentators, the unique cases sometimes quoted are highly suspect as rustic borrowings, as even a glance at the more primitive Anglo-Saxon evidence indicates.[25] That both the sun and the moon were still respected as ultimate sources of power, and particularly growth, is not to be denied. They might be feared, or even propitiated, but neither was worshipped. Both had long been assimilated into Christian lore or the church calendar. Indeed, perhaps the most marked feature of the folkloric evidence at this period is the absence of direct information conclusively linking, for example, the sun with the beneficial qualities of fire. If there was such a link in people's minds it has now to be inferred from the timing of ritual fires at Christmas (after the winter solstice), from the associations connected with the wood used, or from the prevalence of red or gold as colour symbols on certain occasions. It may thus be suspected that in many such rituals an element of convention had long been present.

There can be no doubt that a belief in the existence of forces abroad in the terrestrial world, however it originated, lay behind a mass of superstitions, rituals and popular cures. However we term them, such forces were deemed to be neither good nor bad, neither necessarily dominant nor invariably submissive to more powerful agents. All depended on the use to which they were put. Thus, to take but one example, a widespread superstition associated unusual hares with the physical transformation of witches; yet a hare's paw hung around the neck of a sufferer, or kept in his pocket, might be used as a specific against rheumatism or cramp. The connection was presumably that both afflictions impeded swift bodily movement.[26]

It was clearly because of this potential ambivalence that so much emphasis was placed on the correct observance of custom. In Herefordshire, for instance, the last sheaf cut at harvest appears to have been thought potentially so dangerous that the men threw their sickles at it from a distance – even over their shoulders – until by chance the stalks were cut.[27] The significance of the ubiquitous corn 'dolly' made out of this last sheaf thus probably lay not in its form, which anyway varied, but in the fact that it was plaited before being brought into the house. The same was true of other types of vegetation used for ritualistic purposes: the ashen faggot or the hawthorn bush at Christmas, even witch-sticks of rowan.[28] All were wrought, bound, or knotted. It is relevant to note that superstitions frequently prohibited either the picking or the bringing into the house of wild or even garden flowers, except for culinary or medicinal purposes.[29]

The essence of popular attitudes, in fact, lay in a medieval desire for balance and harmony. There was a proper place for everything. Whenever order was disturbed in one sphere a disturbance might occur in another, as in the common jingle.

> A Saturday's moon with Sunday full,
> Was never good and never wull.[30]

Moon-day should properly follow Sun-day, which therefore ought to be given primacy. After all, the Man in the Moon was banished there for working on Sunday.[31] Troubled times for society would follow for the whole month if a new moon appeared not on the third, but on the fourth day.[32] Two moons in the already dangerous month of May meant rain throughout the entire growing season, for by definition a month should contain only one new moon.[33] An abnormality in nature thus implied a disharmony elsewhere which had to be interpreted. A snake on the doorstep, a bird tapping at the window – or, even worse, flying into the house – was so unnatural a trespass by a wild creature that death, illness, or at least misfortune in society would be portended.[34]

Consequently, the future, not the past, was what mattered. History was usually relevant only in so far as legend might account for an unusual hillock, a prehistoric barrow, or a stone circle in terms which perpetuated the structure of superstition.[35] What did concern traditional country people was how to anticipate the future from signs provided in the present, whether they were indications of the arrival of a stranger, love divinations, weather omens (of the 'red sky at night' variety) or the portents of misfortune. Apparitions of persons about to die (even one's own neighbours) might be watched for in the yard or porch of the church on St Mark's eve (24 April) or All Saints.[36] Only in some cases was avoiding action possible. On different parts of the coast, for instance, fishermen would refuse to put to sea if they saw a hare, a rabbit, a pig, or a priest while going to their boats.[37] For the rest, a degree of fatalism was implicit in such attitudes and it complemented the wretched living and working conditions in which these people usually existed.

Even if unanticipated, there were thus many ways in which current misfortune could be explained retrospectively. A palsied hand might be accounted for by the fact that it had been instrumental in breaking a taboo, such as killing a robin.[38] Similarly, a repeated moral of numerous folk-tales was that bad luck fell on those who failed to observe correct rituals like propitiating the fairies with a dish of cream, or committed ritual sacrilege by felling particular trees, for example.[39] When explanations of this kind were exhausted there was still an ultimate scapegoat available in the person of the witch and hence, hopefully, remedies to be found on the advice of a cunning man.[40]

The country dweller thus sought to relate the fortunes of society to the signs and manifestations of the supernatural. Above all he evaluated nature not simply in terms of practical needs like food or fuel, but also in its living relation to culture. Within nature a tension was seen to exist between those forces that might be used on behalf of society and those that could be used against it. On the one hand were those aspects of nature which were most usually endowed with a supernatural quality of a non-malevolent kind: certain trees whose height and/or age separated them from mankind, and whose wood or foliage furnished a protection against witches; certain plants like the house-leek which when planted on the roof preserved a house from lightning; certain winged creatures, whether especially sacred birds like robins or wrens or various groups of birds (like magpies or rooks) whose number or behaviour might indicate death (but not cause it), and even bees; and, finally, certain subterranean mineral substances such as running spring water, iron, and silver – even coal – all of which could be employed in emergency as antidotes to the most threatening manifestations of the witch. On the other hand were those creatures – often nocturnal – that occupied much the same terrestrial plane as society and which were associated with the physical transformation of witches when secretly attempting to milk the cows, hares, hedgehogs, bats, and even nightjars which nest on the ground.

In the contemporary folklore of creatures and plants, indeed, there were few living things that were not connected in one way or another (if only as popular cures) with the supernatural forces with which nature was held to be imbued. Though the evidence of these specific matters is embarrassingly rich, it must be emphasized that we cannot now tell, and perhaps never may, how far such beliefs dominated the day-to-day thoughts of the country dweller in early Victorian England. It seems possible that increasingly he or she turned to such expedients only *in extremis* or when convention dictated.

The decline of the pre-scientific attitudes of mind which have been broadly outlined here is extraordinarily difficult to chart precisely. Not only has the subject as a whole been pointedly ignored by most historians, but the evidence is such that it is hard to disentangle an unambiguous general trend for the equally unambiguous indications of pockets of local resistance. Even the pace and timing of the change is in doubt. There are no censuses of mental attitudes to help us. In this context, and in the space available, it is thus only possible to propose three imprecise measures with which to assess the process.

The most intangible is that which relates to the altered structure of beliefs. It has already been suggested that convention may well have marked at least certain calendrical rituals early in the century. It may now be added that different superstitions relating to the same natural objects frequently range over such a spectrum of different significances that it may well be wondered how far any coherence behind such beliefs (if indeed that ever existed) was still perceived. When all allowances are made for local diversity, the implication must be that attitudes were increasingly fragmented. One family may have passed on one belief; another a quite different superstition relating to the same object. It is worth emphasizing, none the less, that even one such superstition, sincerely believed, could still have meant a passing attachment to that wider framework of attitudes which had once existed.

A second factor to be considered in assessing cultural decay during the later nineteenth century is more concrete. What is most probably at issue is the decline in the numbers and influence of the people available to hold these attitudes in common rather than the rapid dissolution of the beliefs themselves. For this situation, and important as it no doubt was, rural depopulation provides only part of the explanation. There was also a marked shift in the class, sex, and age-groups of those who perpetuated the old attitudes. Where it had not occurred already, labouring families, and especially the women and children (who everywhere now dominated May Day, for example), were rapidly becoming the sole repositories of local customs.[41]

The third approach to the problem of decline has consequently to concentrate on the circumstances in which all these beliefs were both fragmented and perpetuated by fewer people. No single factor can be held responsible. The probable causes range from the increased activities of church and chapel through to the educational advances of the last decades of the century; from the incarceration of impoverished old women – potential witches – in the workhouses of the New Poor Law, to the increased experience of town life; from the growing mechanization of arable agriculture to the advent of cooking stoves in areas like the Lake District, where the old significance of the open hearth was consequently diminished; from the spread of friendly societies, and their attendant ceremonies, to the development of agricultural shows with their emphasis on more scientific farming. These and many other reasons may be advanced. What is more certain is that following the geographical and social intermixing engendered by the First World War, the year 1918 – unlike 1851 with which this survey began – finally and symbolically clanged shut, like a blood-stained cast-iron gate, on an already dwindling cultural tradition that had evolved in one form or another over more than a thousand years.[42]

## Notes

1 Henderson, 1866, 244–5.
2 ibid., 221–2; Simpson, 1973, 73–4; Briggs, 1971, II, 685.
3 Atkinson, 1891, 110–25; Leather, 1970, 53.
4 Torr, 1918, 8.
5 Atkinson, 1891, 73.
6 Plomer, 1964, 137.
7 Simpson, 1973, 75.
8 Courtney, 1890, 142.
9 Hole, 1944–5, 120.
10 Rollinson, 1974, 78; Courtney, 1890, 141; cf. Evans, 1966, 160.
11 E.g. Henderson, 1866, 264, 140–4; Gurdon, 1893, 20–1; Leather, 1970, 82–3.
12 Torr, 1918, 7.
13 Evans, 1970, 17.
14 Cawte, Helm, and Peacock, 1967, *passim*.
15 Evans, 1966, 18–19 *et passim*: Simpson, 1973, 77–8.
16 Howkins, 1973, 5.
17 Ellis, 1889, *passim*; Wakelin, 1972, 48–51, 102.
18 Grigson, 1975, 290–1, 146–7.
19 Northall, 1892, 8.
20 J. H. (mine-engine driver, 1871–1908), 'Reminiscence of Old Times', dated 1930, 12. Transcript kindly provided by K. C. Phillips.
21 Henderson, 1866, 123–5; Billson, 1895, 35; Simpson, 1973, 148; Killip, 1975, 184–6; J. G. Jenkins, 1976, 141; Opie and Opie, 1959, 288–9.
22 Phythian-Adams, 1975, 10–30; Malcolmson, 1973, 118–38.
23 Evans, 1966, *passim*.
24 Briggs, 1970, I, 135 *et passim*; Briggs, 1971, II, 620 *et passim*.
25 Phythian-Adams, 1975, 9–10; Baker, 1974, 23–4; cf. Leather, 1970, 88.
26 E.g. Briggs, 1971, II, 642–4, 664–7, 699–700, 704, 736–7; Evans and Thomson, 1972, 160–77, 234; Henderson, 1866, 201.
27 Leather, 1970, 104; cf. Jones-Baker, 1977, 156–7.
28 Minchinton, 1975, 71; Leather, 1970, 91–3 and plate; Hole, 1944–5, 90–1, figs.
29 Baker, 1974, 62–3, 84.
30 Northall, 1892, 462; Baker, 1974, 23; cf. Henderson, 1866, 114; Courtney, 1890, 136; Bilson, 1895, 146.
31 Briggs, 1970, I, 123; cf. e.g., Simpson, 1973, 30.
32 Dack, 1911, n.p., 'Folk Lore (3)'.
33 Briggs, 1970, I, 543–5.
34 Leather, 1970, 119; Baker, 1974, 64; Henderson, 1866, 49; Dack, 1911, n.p., 'Birds'.

35 Jefferies, 1880, II, 204, Grinsell, 1976, *passim*; Briggs, 1971, II, 151–400.
36 Gurdon, 1893, 32; Henderson, 1866, 51–3; cf. Billson, 1895, 61.
37 Courtney, 1890, 135; Evans and Thomson, 1972, 234; Simpson, 1973, 153; Killip, 1975, 122–3; Jenkins, 1976, 182.
38 Gurdon, 1893, 7; Courtney, 1890, 163.
39 Briggs, 1970, I, 240–5, 439–40.
40 Courtney, 1890, 139.
41 E.g. Howkins, 1973, 13, 14; Flora Thompson, 1939, ch. XIII; Ruddock, 1964–5, 69; Opie and Opie, 1959, 1–5, 232–92.
42 Evans, 1970, 103–5; Evans, 1966, 17–19; Howkins, 1973, 64.

6

# Voices from the past: rural Kent at the close of an era

Michael Winstanley

One point of view is seriously under-represented in the tomes of evidence bequeathed to us by our Victorian predecessors – that of the humble labourer and his family. We are generally obliged to evaluate their way of life through the eyes of others: employers, officials, their 'betters'. The reason for this dearth of information is not hard to find. Even those men and women who had the ability to leave us their impressions rarely considered their simple lives worthy of recording for posterity. 'I feel that I have not done much that is worthy to be left on record,' commented one Kentish labourer who did try to write his autobiography in 1912.[1] Consequently he concentrates on the exceptional rather than the everyday occurrences, on detailed descriptions of places he remembers and on the personalities of local gentry with whom he came into contact. Many questions we would like to ask are left unanswered.

Fortunately, within the last few decades we have developed both the facility and desire to tap these reservoirs of experience. 'Oral history', the ungainly title given to the process of tape recording recollections, has arrived just in time to allow the last generation of Victorians to leave a more permanent record. The idea of turning to the older generation for knowledge is, of course, far from novel, but the technology which enables us to record it verbatim is. Had George Sturt had access to a tape recorder in the 1900s his attempts to reconstruct 'the old rustic economy of the English peasantry' as it had existed in the mid nineteenth century would have been made much easier and the results more vivid.

> Here at my door people were living in many respects by primitive codes which have now all but disappeared from England. . . . The perception came to me only just in time, for today the opportunities for further observation occur but rarely. The old life is being swiftly obliterated . . . in another ten years' time there will be not much left of the traditional

> life whose crumbling away I have been witnessing during the twenty years that are gone.[2]

Not until the 1950s did George Ewart Evans, using the new technology of the tape recorder, re-examine the 'traditional life' in the same way. Today the passage of time has reduced our period of study still further. We can only just stretch back to the turn of the century and the survivors are few.

The countryman's memories are not concerned with nationally significant events. Queen Victoria's Jubilees, the Boer War, the activities of Lloyd George – these excepted, there are few references to happenings which made the newspaper headlines. Absent, too, are precise dates. The calendar he refers to is that of his own life. His recollections tend to be parochial and personal, dealing with his family, his work, his village, his daily life; but this in no way detracts from their significance, as some sceptics would maintain. Through such details we are better able to understand the values and outlook of our predecessors. These minutiae are interspersed with shrewd comments based on personal observation and experience, which are capable of opening doors of understanding which would otherwise remain locked. Once convinced of the utility of their memories there is nothing most old folk would rather do than talk about them, a trait noticed by many earlier writers. 'Country people have tenacious memories, and the older ones delight in finding a listener to whom they can relate things which they experienced when they were young or which they were told by their parents or grandparents.'[3]

The outline of labouring life which follows relies almost exclusively on memories collected from elderly people in Kent, a county usually referred to as a prosperous area reliant on hops and fruit.[4] Most contemporaries tended to assume that the labourers were correspondingly richer than in other areas. 'It is a well nurtured land, and the people are a well nurtured people', wrote Richard Heath.[5] 'The Kentish labourer is decidedly better off as a rule than agricultural labourers generally are in some counties I have visited', remarked Aubrey Spencer in agreement in 1893.[6] Even Alfred Simmons[7] admitted that the labourers were 'favourably situated', and most farmers echoed his sentiments, some arguing vehemently that 'he has got too big a share of the cake'.[8]

Wages, however, varied considerably from district to district and even between farms, and it would be wrong to assume that all farmworkers were comfortably situated. Investigations tended to concentrate on the hop and fruit districts where there were opportunities for piecework and plenty of seasonal jobs for women and children. Elsewhere, however, especially in the Weald (affectionately referred to as 'Yellow Belly Country') wages were lower, and on the chalk downs – sheep and corn country – work for the rest of the family was 'irregular and uncertain'.[9] A labourer's age and health could also influence his income. Those then just beginning their working life remember their youth as a time of plenty. 'I was single then. Like the song says, when I was single my pockets they jingled, I wish I was single again.'[10] Married men had less opportunity to save: 'Not a gay living for a large family, 16s 6d per week, 2s for rent, 2s for fuel, 10d for school, myself and wife and 7 children to live . . . I must leave you to judge my condition.'[11] Even for the relatively secure yearly workers, as old age crept on 'every farm lane led to the distant workhouse'.[12]

Throughout the county, therefore, some degree of self-sufficiency was still vital for the labourer if he was to be able to maintain his family. Some non-labourers envied the labourer's apparent ability to supply himself with ample food:

> They lived like fighting cocks. It's true! No question about it at all. They had more than they wanted to eat, at practically no cost at all to them theirselves. . . . Now each of those cottages had, I should think . . . the best part of an acre of ground. Anyway sufficient for each of them to have a little bit of orchard in the bottom. Apples, pears, plums, plus two or three dozen gooseberry bushes, plus plenty of ground to grow all the vegetables they could ever want. They had free milk. No rent to pay.[13]

The labourers and their families are less ecstatic in their views. 'Half the time they [the parents] hadn't got the money to buy seeds to plant with. Might have been out to work and come home and hadn't had nothing to eat. And nothing to eat when they got home. No dinner. No nothing.'[14] Nor did many of them view gardening as a pleasant pastime. Labourers relied on easily cultivated plants which required little attention – cabbages, onions, and especially potatoes, enough 'to last till taters come again'. The uncommon fondness for vegetables, especially soggy cabbage, which Richard Jefferies noted, was the result of necessity not choice. Whenever possible children were 'encouraged' to help in the garden:

> They didn't exactly *make* us, but they encouraged us. If we didn't do so much, well, we'd get a clip round the ear'ole. . . . If we see there was some weeds in the garden, did it voluntary, we might get a penny or a ha'penny at the end of the week.[15]

Despite the abundance of fruit in the county labourers had to rely almost exclusively on their own supplies:

> You could never buy any. They didn't sell them in the village. They all went up to the London markets. Unless you'd got a tree in your garden, then you never had any of those fruits. That's why the children used to pinch fruit. Apples you could buy, but only the drops. All the picked

> ones were sent to London. They say Kent is the garden of England, but in those days there wasn't a lot of fruit about. I always remember that.[16]

Even more of a problem was meat. Butchers' meat, although relatively cheap after frozen imports had begun to arrive, still remained a luxury. 'Mother never let us go hungry. We always had something to eat. The only thing is, we never used to get much meat.'[17] If anyone in the family regularly ate butchers' meat it was the father. The main sources of meat, however, came from outside the market economy. Cottagers either reared their own or found 'other ways' of supplementing their supply. Pork featured prominently in many diets:

> That was the principal living, pork, in them days, you know. When I was as I say, in plough service, I lived on fat pork for twelve month. That's all we had. Bar Sundays. Used to have a beef pudding. But that's all I had for breakfast, dinner and tea. Fat pork. Fairly near done me in you know. Going straight from home too. . . . I don't know if I could stick it now. Cor blimey! And sometimes it wasn't done properly. When the knife went through you could hear it sort of crunch.[18]

So heavily did some cottagers rely on pigs for meat that many of the children were put off it for the rest of their lives, and now refuse to eat it. 'But you can understand that we got sick of pork . . . for years I've never touched a piece of pork. I don't even look at a piece.'[19] A Christmas gift from a local farmer in later years remained uncooked. None of the family could face it.

Not every cottage, however, kept a pig. As well as an agreeable farmer or landlord, it required a certain amount of capital, or a friendly working relationship with the local butcher and miller. In one village the butcher's practice was to donate two weaners to a labourer and supply him with sufficient food to fatten them up, then to repossess one to be sold in the shop and allow the labourer to keep the other. Practices varied considerably:

> My old gran'dad used to say, 'It's no good a man trying to keep one pig. You want three. When they're all fed, one to kill for the house, one to sell to pay the miller, and one to buy three more pigs.' And that's the way they went on and made a do of it.[20]

Most men tried to do it as cheaply as possible, buying the smallest pig in a litter (known variously throughout Kent as the darling, Dannull pig, or Anthony pig) from a local farmer for a few shillings, relying on scraps from the kitchen and from neighbours, only filling him out with barley meal in the final stages. Fear of grain stealing, however, meant that some farmers refused to allow their tied men to keep pigs, and the men lost a valuable source of food as well as a certain peace of mind. As Cobbett had astutely pointed out, the sight of a couple of flitches of bacon upon the rack 'tends to keep a man from poaching and stealing than whole volumes of penal statutes. . . . They are great softeners of the temper and promoters of domestic harmony.'[21] With salted sides of bacon hung around the living room what need had the home of artificial decorations: 'You've got pictures!'[22]

Apart from pigs, the labourers frequently resorted to breeding chickens or even rabbits, an additional recommendation in favour of the latter being that their skins could be sold to travelling pedlars. Rabbit stew with swimmers (dumplings) is remembered as a common meal. A few enterprising men even kept the odd goat tethered on the roadside to utilize the free grazing on the verges. In almost every instance the feeding of domestic animals was entrusted to the wife or, even more likely, to one of the children.

These legitimate meat supplies were added to in a variety of ways. Most people from labouring backgrounds recall that much of it was either 'pinched or poached'.[23] 'You could earn a bit other ways. Only you've got to be dumb and simple like the others. But mum's the word. That was the only way you could get through.' Some farmers allowed their men to catch rabbits on their land, but if a pheasant was stupid enough to 'fall out of a tree' and get entangled in the snares then 'we carried him home – never said nothing to nobody – and picked him and burnt the feathers and ate the pheasant. Nobody knew anything about it. We had to be a bit dishonest to keep ourselves going.'[24]

Sparrow catching, on the other hand, was welcomed by farmers; they frequently made generous donations to the rat and sparrow clubs which flourished in many villages, usually with the pub as their base. The sparrow, this 'Avian Rat', was viewed as an unmitigated evil. 'Indeed I do not understand', wrote Rider Haggard in 1898, 'how it comes about that we are not entirely eaten up with these mischievous birds.'[25] For the patient labourer sparrows provided a source of food and, through the local club, a potential, if small, income. Most of them were caught with nets of some description, once again by the younger members of the family:

> We used to go round of a night with a net on two long poles. They used to beat the hedge and we'd be round the other side to slap the net together and put it on the ground. Round the eaves of some of the old buildings you used to have a long stick. The net was on eight-foot woodwork. Used to hold it up and touch the tiles and out would come two or three starlings or sparrows. Down with the net and we'd have them. They used to squawk. . . . Used to get round some of the laurels

> too, round the houses. Toffs' houses and that. . . . Cor, it used to be a game. We used to enjoy it.[26]

Even blackbirds, recognized as useful garden birds, were eaten:

> She [mother] used to pick them like a chicken, truss them all up, and fasten a bit of thread to their legs, stick a fork under the mantelpiece and hang them in front of the old kitchener. Put a plate down the bottom to catch the mess. The heat used to kind of turn these birds, and when it came near to the time they was done, they wasn't half lovely you know.[27]

A thrush was too bitter to eat, but rooks, pigeons and even moorhens were assiduously caught, prepared, and eaten, either in pies or as table birds. A few families were even known to catch and eat hedgehogs. By rolling them in clay and baking them they were able to pull off the spines and skin. 'They used to say they were beautiful you know. But I should never fancy one. Would you?' 'I liked it. It was only the one time I ever did taste it but I liked it then. Of course, we were young, well you'd eat anything in them days, because you didn't get a lot of meat.'[28] Whatever the meat, it was often encased in a pudding or a pie to make it go further. 'You lived on puddin's. Flour and everything would be cheap wouldn't it?'[29]

Self-help, born of necessity rather than moral indoctrination, carried itself over into other areas of life. Children's clothes were invariably home-made, usually cut down from their parents' or some generous donor's cast-offs. Nor was there anything shameful about patches on cloths, although these often symbolized the division between labourers' children and those from wealthier backgrounds:

> You take the butcher's sons and the grocer's sons, publican's sons, that I went to school with, they would all be better off and better dressed than me. . . . Of course the schoolmaster always favoured them people what had got the best clothes on. Us ragged ones always thought they got best treatment.[30]

The skill of a labourer's wife with a needle was all important to the family economy. Few clothes, except the man's and the footwear, were bought. Boots were rarely taken to the 'snob' or cobbler to be mended. Father did them at home. For fuel wood was preferred since it could often be collected by the children. Coal, an expensive luxury, was sparingly used. Monday's washing was done in water heated by wood fires. In almost every aspect of material welfare the labourer's family sought to maintain themselves without recourse to spending their meagre income.

Nothing was wasted. No one was allowed to leave food uneaten. No part of a pig was useless: the 'chidlings' or intestines were scrubbed and used for sausage skins; 'fleed cakes' were made from the fat;[31] trotters, tail, and head were boiled down for brawn and stews; even the bladder found a use as a football or a storage container for lard. Water, a valuable commodity in many villages when it had to be laboriously carried from the pump, was carefully rationed, and sometimes thrown on the garden after it had served its domestic purpose. Tea, despite its dramatic fall in price from the 1880s, remained relatively expensive and was treated as a luxury:

> The teapot was always kept on the hob and the boiling water was always poured into that teapot until there was no more tea came out of it. What came out was water, then that was thrown away. The tea was made in the morning and lasted all day until teatime, and they made fresh tea. I think tea must have been very expensive. You didn't buy a quarter or anything like you do now. It was a pennyworth or twopenny-worth, so it had to last a long time.[32]

Paraffin and candles were conserved by the family's going to bed early. 'Nine o'clock all the street was in darkness. You wouldn't see a light anywhere. People used to go to bed very early. I do now. I always have done.'[33] Old clothes were cut up for rag mats and even rabbit skins were stretched, treated with liquid alum 'till the skin was nice and cured', and then sewn together to make a rug. 'They was quite warm to put your feet on when you got out of bed in the morning.'[34] Old calendars from the village shop were cut up and the pictures used to decorate the walls. Anything that was not eaten or recycled was burnt or buried. There was little need for a refuse disposal service.

Possibly the best-fed labourers, but the worst paid in monetary terms, were the men who boarded with their employer, a practice which lingered on in Kent, especially outside the hop-growing areas, right up to 1914. Mates who lodged with the wagoner were less likely to fare so well. The allowance from the farmer, about 9s. by the turn of the century, which the wagoner had for feeding his lodger was all too often devoted to improving his own family's diet. Several men, however, have fond memories of eating at their employer's table, even if their diet did seem monotonous:

> When we lived in with the farmer, he always had a barrel of beer in, and we used to have small beer for breakfast, ale for dinner and small beer for tea. . . . The only time we had hot tea was on Sunday. For breakfast in the morning, living with the farmer, was a darn great lump of fat pork, half a loaf of bread and, if they'd got plenty of milk, a jolly great bowl of bread and milk. That was something to go to work on. We lived jolly well in them days. After I left that last farm, when I joined the service at nineteen, I weighed 12 stone, 4 pounds. I hadn't

> been in the service three months and I went down to less than 10 stone.[35]

The disadvantage of living-in, however, was that it tended to destroy a man's self-respect; and the achievement of self-respect was just as important as maintaining physical welfare. As a lodger 'you was tied down like a prisoner more or less. A prisoner free. Got it that way then.'[36] How a labourer maintained himself was just as important to him as the standard of living he achieved.

Charity, as an aid to individual efforts, was welcomed and carried no sense of disgrace or degradation, but to turn to the parish was to lose all hope, all sense of decency, all self-respect. It was the source, not the content or level of relief, which mattered. Although 'thousands of our poor village folk seem unable to conceive any greater virtue in their superiors than open-handedness, and nothing but satisfaction is felt on either side',[37] the Poor Law had successfully ingrained in them a dread of parish relief. 'I suppose they thought that's a gift [charity]. That's free. But they wouldn't apply, if it was to do with the workhouse. They'd got that in their heads – the workhouse was a disgrace.'[38] The legacy lingers on even today:

> I grew up with that dread. It was planted in us – save something. You must save so that you don't, in your old age, have to go to the workhouse. . . . Well it's so ingrained in me, that if I was starving, I wouldn't ask for public assistance, because with that it seems the same to me. I know they say it isn't, but anyway I've managed to save enough to do without it.[39]

Respectability also required the keeping of the Sabbath, but here an ambivalent attitude had developed. On the one hand, most forms of work were to be avoided on Sunday. Those who flouted this rule were widely condemned:

> Men out in the country wouldn't dig their gardens on a Sunday. . . . I can remember when I was a boy, there was a man come up to help on the farm and he went out in his garden digging and they talked – I heard my father and them say – they was all alike, the men round that way, 'I don't know who he is!' And they shunned that man because he worked on a Sunday.[40]

Church attendance, on the other hand, was far from obligatory in this moral code. There was no joy in being denounced from the pulpit (personally in some cases), in squeezing into the cold, draughty pews at the rear of the church, in being inspected by employers on the one day off. Not to mouth acceptance of religious mores was looked down on, but church attendance was not, at least among the labouring class, much to the dismay of the parsons. 'If people didn't go to church he'd [the vicar] go round to their houses and give them a dressing down. I don't know whether it made much difference or not.'[41] Although children attended Sunday School, many of them were astute enough to realise their parents had ulterior motives in sending them. 'It was just to get rid of us. But all the kids liked going,' even though they were taught by ' "old toffs' " daughters'.[42]

Status, acceptance by their own kind, was of supreme importance to most families, and their quest for it detracted from any serious questioning of the structure of rural society. Although perpetually conscious of their class, even cynically critical of their social superiors, few labourers seem to have translated this into open hostility. They did not seek to better themselves by attacking their employers' privileged positions, and they brought their children up in the same way of thinking.

> There were certain people who were our superiors and we were supposed to adopt a courteous attitude to them. Nothing really terrible about it. We looked up to them and we expected them to be better than we were, and it gave us some standard.[43]

> You didn't know anything about the haves and have-nots. . . . We wasn't educated to that extent. Them was the days when you lifted your cap when you met anyone.[44]

Social superiors who failed to line up to the standards set for them were criticized for their personal flaws, not as typical specimens of the employing class. Generally, even these failings were overlooked because 'they give us a living didn't they, you see?'[45] As Sturt observed:

> This village looks up to those who control wealth as if they were the sources of it: and if there is a little dislike of some of them personally, there has so far appeared but little bitterness of feeling against them as a class. . . . Being born to poverty and the labouring life, they accept the position as if it were entirely natural . . . They suppose that it takes all sorts to make a world and since they are of the labouring sort, they must make the best of it.[46]

The survivors of the period agree. They accepted their position 'as a matter of course'.

It was within their own class, therefore, that labourers and their families strove to attain social acceptability and esteem. Poverty itself was no disgrace if brought about by misfortune, but if the cause was considered to be flaws of character then the family were outcasts. 'That class of people,' as they were labelled, who refused to fend for themselves, to show themselves willing to 'make do and mend', were given little sympathy. They were frequently considered to be beyond reform, defective in personality not unfortunate in circumstances. 'The more they had the more they wanted.' Quite unconsciously the labourers assumed an individualist philosophy, accrediting everyone with the ability, but not always the will, to maintain themselves independently.

Slovenliness or squalor were similarly degrading. To be clean was to be respectable. Dirty people 'sort of got looked down on, because you can keep clean even if you're poor can't you? It doesn't cost much to, although it was a lot more difficult in those days when every drop of water had to be heated for washing and baths.'[47] The contrast with townspeople, especially Londoners, struck one lady who, from 1900, spent several years working for some sisters of mercy at Rotherhithe:

> You could always depend on country people being clean couldn't you? You couldn't town people. If you went into their houses, when you went back home, you knew you'd got to strip and have a bath, and put all clean clothes on because you never knew what you picked up in their houses. I think country people would have died rather than not be clean. I know when we were children, we played in the dirt, we got filthy dirty, but we never went to bed dirty. We were always washed and clean before we went to bed. And that was clean dirt to what I got used to in London. . . . I can't describe the smell. It was terrible. . . . This smell in London would stick. Their houses used to stink and they couldn't smell it. They were so used to it. It was a dreadful smell. . . . Very different from country people. The difference was dreadful to me because I had always seen my mother cooking and cleaning and making, but not these people. They would sit for hours outside their doors talking and gossiping, but not to do any cleaning. Very few people were clean. . . . I don't think you would understand what the dirt meant to me. I used to shudder.[48]

Many Kentish people had no need to travel to London to compare themselves with their city counterparts. The city came to them for hop-picking each year, invading their privacy, overrunning their pubs pilfering from their shops, and displaying their filthy habits for all to see. Country dwellers, almost without exception, received a huge stimulus to their self-respect. The hoppers might be 'not too bad', 'tremendously good hop pickers', 'sociable people', and 'all right as a rule', but they were widely regarded as an inferior class of beings. Their women went into pubs and could be seen smoking, both criminal acts in the rural code. Above all they were dirty. 'After they went back we always used to pray for good rain.'[49] The greatest disgrace for a local child was to return to school with his or her head shaved, a victim of the lice which the seasonal migrants brought with them. Disinfectant, diarrhoea mixture, changes of clothes, the burning of straw and bedding left behind by the visitors, all were employed by the labourers to keep themselves free from infection. Everything about the Londoners indicated that they were a class apart. This self-righteous indignation persisted right down to the end of the 1950s when the Londoners, by now much altered in their habits and enjoying more hygienic lodgings, ceased to migrate in such numbers. Until that time, however, the labourer in the hop areas of the county was reminded annually that he was far from being a member of the lowest stratum in society.

Physical welfare and status – but what of that elusive quality of life? Were people actually happier in those days? There is a great temptation today to view the slow, peaceful ways of the horse-powered rural economy as part of a lost golden age. This is not a new phenomenon. Writers have always been at pains to point out that the idyllic days of the old organic community, traditional England, have just, and only just, departed. For Cobbett the golden age had already gone by 1820; for Sturt it was dying only from the 1890s. For the present generation of writers it was alive up to 1914, possibly even up to the 1930s. No doubt future writers will view the present decade with similar affection. Few of those old enough to remember the turn of the century would concur with the golden-age image. Although they bemoan their lost youth and hanker after their 'good young days', they reject out of hand any suggestion that their way of life then was intrinsically better than the one they now enjoy. 'They weren't about then, some of them what talk about it [the good old days]. Not when I was.' As for wishing to see them return: 'Oh God, no! No! Where there was a big family, pretty well starved out we was.' 'We was more or less convicts. You didn't dare smoke or anything like that at work.'[50] At best the labourer lived under a benevolent despotism, at worst a tyranny.

Yet elderly folk would agree that something has been lost in the changes of the past eighty years: a certain contentment. Lacking an ideal, living their lives 'in grooves', as Jefferies astutely called it, they merely desired to make ends meet, an aspiration pitched low enough for the majority of them to achieve it most of the time. If this was attained they were satisfied, they were content, and relatively indifferent to all else around them, a trait which commentators like Heath attributed to their 'unceasing, protracted labour' which robbed them of any chance to aspire to higher things. As many recent social investigators have pointed out, perceptions and expectations play a significant role, possibly a more significant role than actual physical welfare, in determining an individual's level of satisfaction with his position in life. Most labourers recognize these two aspects and deliberately choose the words 'contented' or 'satisfied' to describe their outlook, in preference to the emotive, subjective 'happy'. 'They was hard old days but as I say they were peaceful. People knew that was their lot and they wasn't going to get no more. But nobody's satisfied today. The more money they got, the more money they seem to want.' 'Oh, things have improved in ever so many ways. Don't think I want to go back to my day – I don't. . . . Yet we was happy you know. We was contented, because we didn't know any different. We didn't expect anything, because if you did you didn't get it.'[51] Comments like these are not exceptional. They are typical of the majority of people who chose to remain on the land, and, surprisingly enough, many town dwellers express similar opinions. As children and young adults they were subjected to social forces which made it only too plain to them what their lot in life ought to be.

Today we tend to consider that Victorian labouring families lived in what we would call poverty, but only with hindsight and a great social awareness have many of those brought up in such conditions come to perceive their position as having been in any way unnatural, unsatisfactory, or resembling poverty:

> We never thought we were poor then, but we were. . . . I mean you never thought about it, but afterwards when you think how you've been exploited and have been exploited over the years, that's when you think about it. But not then. It never entered your head. It was just right that it should be so. . . . You never thought you had a *right* to have anything.[52]

> We were all poor. Well I consider we were very poor, but it's only late years I've thought we were poor. . . . There was enough to eat, and just enough. . . . I don't think you can realise how very poor people were, and yet to me they were always quite happy. There was no grumbling – I never heard anybody grumble. We could always manage to get something to eat.[53]

This 'self contained oblivion' or 'Oriental absence of aspiration'[54] which Jefferies found so depressing is the key to understanding the labourer's 'contented indifference', an attitude rudely disturbed by four years of war from 1914 and challenged over the years by the development of media and advertising.

What remains today? Agricultural methods have undergone another revolution. Farm cottages are now second homes or occupied by town workers eager to 'get away from it all'. The world, as the survivors of the Victorian age remember it, has vanished. Yet these survivors remain virtually unchanged, firmly embedded in the past, still sitting in the rear pews of the church where they were taught to sit as children, wary of drawing social security, the modern parish relief, avoiding pork, the staple of so many of their earlier meals. They are reminders that customs and values linger on for a long time after they have ceased to be meaningful. Sturt had looked forward to 'a renaissance of the English countryside', but those who had little or nothing to look forward to then now have even less. They have only the past. Their memories are their only real possessions.

## Notes

1 T. J. Pointer, 'An agricultural labourer's autobiography' (1837–1912), MS Broadstairs Public Library.
2 Sturt, 1912, 8, 17.
3 Cornish, 1939, 117.
4 Extracts are from a few of the interviews collected between 1974 and 1977 for the SSRC project. 'Life in Kent before 1914'. Page references are to transcripts of interviews. All the respondents were born in the last two decades of the nineteenth century and come from labouring families unless otherwise stated. They make little distinction in their recollections between the late Victorian and Edwardian eras: the First World War is more of a watershed for them.
5 R. Heath, 1893, 160.
6 BPP 1893–4 XXXV, 665.
7 Secretary of the Kent and Sussex Labourers' Union, BPP 1882 XIV, 170.
8 Charles Hoare, Staplehurst farmer, BPP 1894 XVI pt 2, 503.
9 BPP 1868–9, XIII, 81.
10 Albert Patterson (born 1892), 18.
11 BPP 1882 XV, 165.
12 Green, 1920, 108.
13 P. C. Robinson (born 1892), bailiff's son, 12.
14 Freddie Moon (born 1887), 14.
15 Jack Larkin (born 1889), 15.
16 Bernice Baker (born 1884), 39–40. See also E. Harrison, 1928, 253.
17 William Darby (born 1892), 16.
18 J. H. Barwick (born 1886), 14–15.
19 Baker, 24–5.
20 Larkin, 18–19.
21 Cobbett, 1821, para. 139.
22 Steve Prebble (born 1891), fisherman/farmer, 27.
23 Patterson, 28.
24 Percy Barnes (born 1889), 21, 36.
25 Haggard, 1906a, 90.
26 Harry Gambrill (born *c.* 1890), traction engine contractor, 42.
27 Gambrill, 44.
28 Barwick, 47; Frank Kemsley (born 1889), horse dealer/higgler/farmer, 20.
29 James Styles (born 1887), country building labourer, 6.
30 Styles, 14, 28.
31 Fleed: the internal fat of a pig from which lard is made.
32 Baker, 75.
33 Baker, 28
34 Barnes, 20.
35 Larkin, 4, 9.
36 Patterson, 7.
37 Bennett, 1914, 42.
38 Baker, 53.
39 Baker, 6.
40 Frederick Atkins (born 1883), labourer/thatcher, 4–5.

41 Barnes, 108.
42 Larkin, 46–7.
43 Harold Pilcher (born 1897), smallholder, 35–6.
44 George Post (born 1896), tailor, 18.
45 Winifred Beech (born 1894), 19.
46 Sturt, 1912, 103, 106.
47 Dora Fenney (born 1892), publican/carter/labourer, 51.
48 Baker, 18–20.
49 Laura Bryant (born 1892), 9.
50 Moon, 22; Barnes, 60.
51 Barnes, 35; Styles, 9, 10.
52 A. Fordred (born 1896), 28.
53 Baker, 2, 32.
54 Jefferies, 1880, II, 78.

# 7

# The farmers in the twentieth century

B. A. Holderness

In some sense the farmers of Britain have endured the vicissitudes of prosperity and depression and the distractive force of technological change during the twentieth century better than other social groups in agriculture. For two-thirds of the century even the decline of numbers in the profession showed little acceleration from late nineteenth-century trends. After the mid 1960s the fall was steeper, encouraged by government policies of rationalisation, but the proportion of agricultural labour provided by farmers and their sons (or daughters) nevertheless increased, since the shake-out of wage-earners and the recession of the landed estate as a feature of the rural economy were even more pronounced. The desertion of the countryside by those engaged in working the land is reflected in forsaken or obliterated steadings, in farmhouse and barn conversions, as much as in derelict mansions or redeployed cottages, but the perspective of the late twentieth century is pitched upon a short focus. By comparison with the past twenty-five years the erosion of agrarian manpower was almost geological in its pace between 1840 and 1960.

Measuring the change is difficult. The census of occupations offers less consistency of approach than had been the case between 1851 and 1901, and it becomes especially difficult to use after 1951 when sampling was adopted. M. C. Whitby surveyed the material from four censuses, those of 1921–61. His results were not strictly comparable (table 7.1) but they do tend to confirm the proposition that the numbers of farmers remained buoyant.

The methodology of collection makes it difficult to compare pre-war and post-war data in order to calculate turnover, but inter-censal changes can be analysed. Whitby concluded that in 1921–31 the net rate of entry into the profession was 16.6 per cent and the net rate of withdrawal 24.2 per cent. For 1951–61 the figures are 16.8 per cent and 27 per cent respectively. In both decades there was an actual loss to the industry of over 7.5 per cent. Between 1901 and 1911 there was a net gain in recruit-

**Table 7.1** Farmers in England and Wales, 1901–61

| | *Farmers and managers* | *Market gardeners* | *Bailiffs* |
|---|---|---|---|
| 1901 | 224,299 | n.a. | 22,662 |
| 1911 | 228,788 | n.a. | 22,166 |
| 1921 | 264,093 | n.a. | 22,479 |
| 1931 | 248,246 | n.a. | 9,847 |
| 1951 | 275,525 | 51,385 | 15,446 |
| 1961 | | 298,296 | |

Sources: 1921–61; M. C. Whitby, 'Farmers in England and Wales, 1921–61; *Farm Economist*, XI, 1966, 84; 1901, 1911, *Census of England and Wales, 1911*, (C.7018), vol. X, pp. XIV-XIX.
Note: returns of market gardeners are given only for 1951, because the census data otherwise include their labourers, in 1901–11, and domestic gardeners, 1921–31.

ment of 1.8 per cent, expressed as the difference between those entering and those leaving agriculture, and in 1911–21 the net gain was as high as 14.2 per cent. What happened between 1931 and 1951 is very uncertain. The inference is that there was a substantial net increase of individuals in the profession, for which the evidence is dubious. The number of holdings in excess of five acres declined between 1931 and 1951 by 16,000 and fell to 248,536 in 1966. From 1967–8, following an initiative by the government after 1966, numbers began to fall more rapidly. In 1971 there were about 270,000 farmers and salaried managers in England and Wales, and in 1981 237,031. Using the Whitby formula we can suggest that between 1951 and 1966 about 170,000 individuals withdrew from agriculture, of whom perhaps 36,000 were available for other work; the remainder who died or retired were replaced by new entrants. In the next fifteen years the net decline exceeded 20 per cent or about 56,000 individuals.[1]

The census is inadequate as a source of information about entrepreneurs in the industry. Farmers recorded in all the returns included both active and retired individuals, the holders of both unitary or joint enterprises, and men and women with other occupations who nevertheless preferred to be thought of as 'farmers'. On the other hand there were occupiers of land in every county, but especially where hobby-farming was prevalent, who opted for a different designation. It is impossible to calculate how the rising status of the profession affected enumeration. Local studies imply that some established landowners (formerly gentry) chose to be called 'farmers' in the 1940s and 1950s, and for many amateurs of husbandry the term also appeared a more attractive sobriquet than auctioneer, butcher, feed merchant, or even civil servant. The census, for all its precise ordering of occupations, therefore leaves much to be desired. The problem is even more complicated by comparison with the annual return of agricultural holdings in the Agricultural Statistics.

Agricultural holdings in excess of five acres are summarized for particular years in table 7.2. Differences in the method of collection were not extensive but one major problem makes the basic data difficult to use. Farms that spread over several 'holdings' were treated inconsistently. Some were classified as single enterprises; others were broken up into their component holdings, often it would seem at the discretion of the farmers making the returns. We need to know how many 'enterprises' existed, rather than how many holdings or how many so-called farmers there were. This was done, in part, by Ashton and Cracknell for 1958–9. They concluded that there were 30,368 multiple holdings which were merged in 13,444 enterprises (out of 358,000 holdings all told). For 1959 they also worked out the legal status of the occupiers of holdings. Out of a total of 357,275, 314,571 were in the hands of individuals, 31,786 in partnerships, and 7,053 in joint-stock companies. On a different calculation the ratio of enterprises to holdings apparently varied between 0.067 and 0.082 between the mid 1950s and the late 1970s, with a tendency for amalgamation to reduce the number of active farmers in business. In addition to the halving of the recorded holdings between 1900 and 1980, hidden engrossment accounted for another 10–15 per cent of losses in personnel. The number of occupiers, in other words, declined from perhaps 400,000 in the late nineteenth century to 160,000 in the 1980s.[2]

**Table 7.2** Agricultural holdings in selected years in England and Wales

| | *5–20 a.* | *20–100 a.* | *100–500 a.* | *over 500 a.* |
|---|---|---|---|---|
| 1895 | 126,714 | 131,637 | 79,775 | 4,523 |
| 1915 | 120,616 | 138,087 | [80,480] | [4.400] |
| 1925 | 110,385 | 140,050 | [76,490] | [3,500] |
| 1935 | 96,882 | 137,372 | 75,253 | 2,997 |
| 1944 | 88,487 | 129,082 | 74,341 | 3,337 |
| 1964 | 71,905 | 110,078 | 70,047 | 4,953 |
| | *5–15 a.* | *15–100 a.* | | |
| 1975 | 26,964 | 89,294 | 62,770 | 7,088 |
| | *2–10 ha.* | *10–40 ha.* | *40–200 ha.* | *over 200 ha.* |
| 1985 | 31,747 | 48,002 | 53,711 | 9,308 |

Source: Agricultural Statistics of the United Kingdom; before 1964 abstracted in Ministry of Agriculture, Fisheries and Food, 1968, *A Century of Agricultural Statistics, 1866–1966.*
Note: Figures in square brackets are estimated.

The figures suggest that a considerable body of men, and a few women, occupied land as a by-employment to other careers. Hobby-farming had been popular for generations and its appeal hardly faltered even after 1970 when the price of land began to rise steeply. Well-to-do business people and semi-retired professionals found farming to be both a relaxation and a tax-efficient means of disposing of surplus profits, particularly after 1945. The current of agrarian romanticism which flowed strongly after 1880 also buoyed up the aspirations of less affluent fugitives from industrial society. Hobby-farming has been a shifting adjunct to agriculture, but where it has been prominent as a feature of rural life, on the outskirts of the greater towns or in half-abandoned upland townships, the effect upon the inherited traditions of farming has proved disturbing or disintegrating. Hobbyists have not necessarily been dilettanti, however; many had both the wit and the means to equip or modernize their holdings as profit-making enterprises, and for several the attraction of offsetting business profits elsewhere led to extensive, expensive, and often effective agricultural investment.[3]

Dual occupation, like the system common before the mid nineteenth century is not extinct. The opportunity to combine two enterprises sympathetically in an integrated business career has remained, because there are still numerous agricultural service occupations compatible with landholding. Auctioneers, butchers, machine contractors, publicans, veterinary surgeons, and agricultural advisers scarcely qualify as hobby-farmers, since for many the occupation of land is collateral to their careers. They are best regarded as contemporary examples of an ancient practice in rural society. This kind of farming has not become systematically organized as in, say, southern Germany, but since throughout this century there have been many farms insufficient to provide full-time incomes for their occupants, opportunities for diversification have often been exploited by countrymen as much as by outsiders. In the 1960s it was found that about half of all farms did not pay as full-time enterprises. On the basis of 'standard man days' (s.m.d.) a holding not providing work for at least 275 s.m.d. per annum was defined as part-time, even though it was obvious that many such holdings were occupied by farmers who had no other work. In England and Wales in 1960, 11 per cent of occupiers of nominally 'part-time' holdings had no other source of income: 42 per cent had other full-time employment, and 14 per cent other part-time employment. The true proportion of part-time holdings in England and Wales in the early 1960s may have been between one-third and two-fifths of all farms (i.e. about 80,000) rather than half, although in Scotland part-time farmers have long occupied at least 45 per cent of the whole number. In the 1980s the number of part-time farmers in Britain as a whole was actually increasing while those in full-time possession of the holdings were in retreat. By 1985 it is estimated that one-third of all farmers were part-time.[4]

Recruitment into farming changed significantly with the dissolution of the great landed estates. For whereas it would not be right to suggest that admission to the mystery was actually controlled by the administrators of estates before 1900, their influence upon opportunities to enter or to make progress in farming had been considerable for several generations. In this century entry into agriculture has been determined differently. With the growth of owner-occupancy, waiting upon vacancies within a tenurial system slow to change has become less significant than the provision of capital for purchase or the accident of inheritance. Owner-occupancy, although it enfeoffed many sitting tenants (not necessarily always to their advantage), was yet less symptomatic of radical social change in the countryside than is implied by the demise of the landed estate. After 1945, and even more strikingly after 1960, the number of farmers entering the profession through purchase was negligible. New money invested in agricultural property originated in business corporations or other institutions, which cast about for partners or tenants to work their land. A large proportion of the land thrown on the market, however, was acquired by families already entrenched in farming. Before 1940 circumstances were different, since the appetite for land in hard times was less keen and many farms were acquired defensively by those uncertain of the future and apprehensive for their posterity. Thus the break-up of estates created opportunities that were different according to the economic prognosis for agriculture – and this obviously affected the ladder of opportunity for individuals.[5]

The pattern is simple enough, but the particular circumstances of each transaction add complications not easy to weave into the basic design. There was some movement towards owner-occupancy before 1914, but it was not equal to the dispersion of settled estates. On the other hand, the turnover of tenancies was especially significant after 1880 when many dynasties of tenants parted with their long-time landlords. From my own collection of estate data it appears that about half of farms not sold changed hands (i.e. passed from one family to another) between 1890 and 1914.[6] The war stabilized the situation, but the profits that accrued before 1918 enabled hundreds of farmers to buy land in the post-war auction of rural Britain. The burdens thus assumed proved troublesome in the 1920s, when land could otherwise be obtained cheaply. In fact, there was a glut in land supply for the decade after 1921 and agricultural freeholds or tenancies remained plentiful in relation to demand until the Second World War. After 1945 scarcity gradually returned and farmers, conscious of prosperity of unprecedented duration, bid up the price of land until by the mid 1970s agricultural property was too expensive for almost everyone but expansionary farmers themselves. The circumstances in which this occurred resembled the land hunger exploited by many estates in the 1850s and 1860s when tenancies were even let by tender. These fluctuations in the demand for land by those who wished to farm it affected the course of recruitment into agriculture, since access varied between easy and difficult in a century when both the trend to owner-occupation and the trend to amalgamation were set.

Whitby's evidence suggests that most changes of occupation were by way of inheritance. New entrants in each decade of the century have typically been farmers' sons. Little research has been done on the background of patterns of inheritance and mobility of farmers in any period for the whole country, although James Nalson's pioneering study of the north-west suggests what can be attempted.[7] Farmers have certainly formed a relatively closed group in the past two generations. A survey in 1944 showed that 81 per cent of the farmers questioned were the sons of farmers, and only 7 per cent were recruited from occupations not directly related to agriculture. Moreover, two-thirds of farmers' children over 14 were either already in farming or intending to enter the profession, though not necessarily as entrepreneurs.[8] Post-war investigations bear out the truth of increasing land-hunger since opportunities for outsiders, except at the top of the profession, seem to have dwindled even further. Of my own sample of farmers, taken in the late 1970s, barely 2 per cent of those who entered farming after 1945 were newcomers.[9] Among their older contemporaries conditions were more variable. Of those who entered before 1939, and were still in business in the 1970s, about one-fifth came from outside agriculture, nearly all drawn out of compatible occupations. Several were the sons of farm labourers, even in Lincolnshire and Norfolk, one was the son of a miner, and another of a railway clerk. There is no correlation between social origin and eventual success; one man who came of a labouring family and had himself begun as a farmworker had acquired holdings totalling over 2,000 acres. Out of the inter-war years a ladder of opportunity certainly led upwards through the prosperity of war and the subsidized peace. These individuals, however, represent success, or at least survival. A third or more of the new entrants into agriculture between 1921 and 1935 failed before the war could rescue their pretensions.

The close connection between farming and its service trades is obvious, but the interaction that led to recruitment into farming also worked in the opposite direction. Farmers, and more particularly farmers' children, passed into trades with which they were already familiar. Little is known about the fate of farmers who failed or sold up before the war. A few, but only a few, ended as farmworkers, at least in their own neighbourhood; imitation of Gabriel Oak was to be avoided if humanly possible. As in the nineteenth century many migrated to the towns and there became lost in the crowd. Anecdotal evidence from my own circle of acquaintance suggests that as many prospered as failed in their new lives, by becoming commercial travellers, skilled craftsmen, and managers or owners of businesses.[10]

After the First World War the expansion of advisory services to agriculture also gave some opportunity to farmers' children. Yet except when farming families were more or less indistinguishable from the rural middle class, the barrier of insufficient education certainly impeded the wider opportunities of the group, particularly before the 1960s. The problem is not small in terms of the rural population. In any decade of this century the number of farmers available for other employment has varied between

10,000 and 30,000 individuals, figures which would be almost trebled if sons and career-minded daughters were included.

At times exclusion from the profession has been enforced by diminishing access to land or through the temptation to benefit from high property prices, but there are other considerations. Tenacious as was the attachment to land in principle, generation after generation, competition among children *not* to assume responsibility for the holding caused much heart-searching on many pre-war farms. Some of those who did bite the bullet were regarded, or thought of themselves, as the least fitted to try their fortunes elsewhere. By the same token, several who had escaped the drudgery of managing barely profitable farms were recalled to their duty, in some instances to take over an unlooked-for acquisition of additional land. The prospect of less isolation or less onerous work appealed as much to farmers and their offspring in the inter-war years as to cottagers. After 1960, by contrast, the problem was often one of finding land or a role within a partnership for sons reluctant to leave the soil. Much of the impulsion to acquire new farms was transmitted through the desire to provide for children. A survey of farming in any part of the contemporary landscape corroborates this drive, since virtually everywhere farming populations are now substantially made up from kin-based partnerships spread across several holdings or from farms separately managed but occupied by members of cognate families. Interrelationships have become so complex, even so dispersed, that the agricultural population often appears more homogeneous than it actually is. In Lincolnshire, for example, one particular clan occupies farms in twenty-one dispersed parishes and is related to four other family groups with about twenty holdings. In such circumstances farming has inevitably developed into an inward-facing profession with an effective method, in the high price of land, of excluding new entrants from outside.[11]

The great capitalists now dominate the industry. Until recently they engrossed much of the land available to purchase or to let. For most of the century they have spoken for the landed interest and conducted its affairs in committees and conference. Subsidies intended to protect agriculture as a whole and to raise small farmers out of the ruck of otherwise irremediably low profitability have perhaps too often fallen to the benefits of the large operators, the rationale of whose business has at the same time been promoted by mass distribution. By the late 1950s the Farm Management Survey showed that the gap in income between the richest and poorest farmers was already wider than it had been in the 1930s. The average income of agriculturists, it is true, had generally improved, from less that £150 p.a. to over £900 p.a. in the twenty-five years after 1935, which was proportionally a better performance than that accomplished by most other economic groups; but since the gap between greatest and least in the survey data had widened from about fourfold to sevenfold, the post-war prosperity did not create a more homogeneous body of landholders.[12]

The differences implied in the income statistics can be seen quite clearly in the rural sociology of the period. The outstanding work on the social milieu and *mentalité* of farmers accomplished by W. M. Williams and his disciples in their studies of villages in Cumberland, Devon, North Wales, and the Cheviots, however, fixed its attention too squarely upon the family farm, even upon a residual peasantry. Such a system, still dominant in the pastoral west before 1970 (and far from extinct in the arable east) was a traditional repository of agrarian wisdom, but even in Williams's day it was doomed.[13] There is obvious poignancy about the changes that have befallen these outlying communities since they were academically inspected. Until Howard Newby hit upon the windy prairies of East Anglia the rural sociology of this mainline agricultural region had been curiously neglected. Newby expressed himself too much in terms of an academic social scientist to capture sympathetically the first-hand experiences of the farming community in the mid twentieth century, but his book, *Green and Pleasant Land?* is invaluable as a work of scholarship and doubly useful as the first of its kind, since he has set the agenda for research into the British farming community.[14] One fact is absolutely clear. Farmers' attitudes, even in some respects their behaviour, have been significantly modified by the changing environment of their work. When it involved a populous hierarchy of employer and employees, often with more or less common experiences shared by both, the farm itself was like a great ship, doubtfully seaworthy, but socially coherent in a recognizable chain of authority. By 1980 the farmer's way of life had become isolated. Farms have now become mere *machines à produire*, and the farmer, looking out across his terrain, is the survivor of a social earthquake. His neighbours, uncomprehending and incomprehensible, are likely to be fugitives from towns rather than farmworkers, blacksmiths, millers, wheelwrights, shopkeepers, or gentlemen 'born to the saddle' who had peopled his village only two generations before.

In literature, the diversity of lives spent in farming has been expressed more completely and perhaps more sympathetically this century than in Victorian writing. The late nineteenth-century rediscovery of rustic virtues and the attempt to recapture the rural tradition in Britain commingled with a new political imperative to engage stalwart countrymen in a struggle against industrialism and its social disorders – in due course also against totalitarian aspirations to reorganize society.[15] The result was a confused mess of discordant ingredients, but at least the literary feast to which the farmers were invited was not disappointing.

On the other hand, it is quite difficult to characterize the position of twentieth-century farming in literature. There remained, even beyond the

Second World War, some sense of the social inferiority with which Robert Martin in *Emma* had been disparaged. Even genteel farmers were made only to touch the lives of gentlefolk, but nevertheless to leave them reeling from the experience.[16] Yet with the decline of landownership as a way of life, husbandry as an avocation, or as a profession, became socially acceptable. Politicians and literary lions were not averse to posing as farmers and even when some went further, like Adrian Bell, and tried to assimilate the spirit and wisdom of old-style farming, refined public opinion was neither outraged nor uninterested at the prospect. In fact, the author Henry Williamson achieved a *succès d'estime* with *The Story of a Norfolk Farm* in 1941. The difference between Bell, who tried to burrow his way into the old rustic certainties, and Williamson, the hobby-farmer, who yet needed and was determined to make his enterprise pay, is instructive. Adrian Bell invested so much intellectual and emotional capital into becoming a 'real' countryman that his literary *oeuvre*, in his *Corduroy* novels, is focused as much on fine detail of country ways, poetic but also practical, as was Gilbert White's vision of the natural history of Selborne. Williamson, on the other hand, undertook his task of turning round a Norfolk farm in the 1930s as an experiment in art as much as in life, and his desertion of the enterprise after a few years gave him no lasting regrets. Both writers set on paper the experience of becoming farmers; neither was born to it, which in itself adds poignancy to their striving at agriculture.

Of the small number of writers in this century who were brought up on farms, very few are of substantial long-term interest, and A. G. Street is perhaps the best example. Our view of a farmer-centred rural universe, relies rather upon writers who used country life either allegorically or polemically. Lewis Grassic Gibbon (Leslie Mitchell), for example, approached British peasant life with deep understanding, and his episodic plots in *Sunset Song* and *Cloud Howe* reveal a deeply felt sense of shared hardship and ancient wisdom unviolated among the poor farmers of the Mearns. Gibbon, like Hardy, is flawed by a dissonant style. His affectation of dialect outside his use of dialogue not only makes the books impenetrable for most English readers but it also gives a spurious ethnicity to his writing which is sentimental when it is not plain silly. He verges in expression on the over-ripe melodrama of Mary Webb, whose Shropshire extravagances justly evoked the satire of *Cold Comfort Farm*. That book, although only a humorous trifle, oddly enough says much that is true about the hayseed mentality of the poor husbandman in pre-war England. Most writers of rural fiction, however, were and are educated townsmen, who like Hardy and D. H. Lawrence allowed an overwrought imagination to transmute base metal into fool's gold. Nine-tenths of books on rustic themes, however good as literature, have misconceived the prosaic character of the farmers' life, often by depicting the English countryman as a species of noble savage.

A. G. Street is the most obvious exception from the middle years of the century. His purpose, beyond story-telling, was to present the reality, often unappealing and always mundane, of rural life. His farmers are real but relatively uninteresting; his plots are pragmatic and plausible; and his point of view is both didactic and explanatory. Street was a first-rate journalist, concerned to place agriculture centre-stage in public affairs, who succeeded as propagandist and man of ideas behind the attempt to modernize and support British agriculture in the 1930s and 1940s. His literary legatees were the authors of the radio serial, *The Archers*, in its palmy days when it concentrated on the life-style of its eponymous farming family.

Imaginative literature may be an uncertain guide to actual rural life but it deserves to be read in conjunction with the rural sociology mentioned above. We cannot otherwise take the measure of the farmers' universe from the mass of official, academic, or journalistic reportage published since investigators first discovered the existence of an 'agricultural problem' in Britain. Treatment of the farmers' interests in most of these outpourings, indeed, has been variously oblique, condescending, or tendentious. Farmers have endured, and not always earned, a bad press, being condemned at first for lack of enterprise or vision and at last for greedy acquisitiveness and a love of new-fangled methods of disturbing the balance of nature. Even in essentially pragmatic works, such as Rider Haggard's *Rural England* (1902), Astor and Rowntree's *British Agriculture* (1938), or the Donaldsons' *Farming in Britain Today*(1967), the tenor of discussion has tended to limit the farmers' point of view and their needs, since to the authors they appear customarily as problems requiring social and political resolution. For every farmer quizzed by Haggard he found two other respondents, and for every farmer with an undeniable tale of success he found another who confirmed Edwardian prejudices about their unfitness for change except by means of radical surgery. By comparison with the intemperate criticism of Garth Christian or Marian Shoard, farmers might count themselves lucky to be so benignly neglected, but it does seem curious that, at least until the 1960s, the central role being assumed by farmers in determining the destiny of the agricultural landscape should not have been fully appreciated by men at the heart of 'rural economy'.[17]

Literary references may easily mislead and if we are to gain a full picture, we must pay attention to the parallel histories of the National Farmers' Union and the Country Landowners' Association. Put briefly, the agricultural interest, when not ravelled up with government, has been represented by farmers for most of the twentieth century. When Colin Campbell and his friends launched the NFU in 1908 they had in mind the needs of men like themselves, practising farmers with capital and some education, who were not so precariously poised as to feel the necessity of aristocratic leadership. In particular, as they confronted the challenge of a Liberal government attracted by land reform, it was essential to establish the point of view of farmers as distinct from that of landowners and land-agents. The disarray of the agricultural interest before 1914 was their opportunity. The old symbiosis between lord and tenant was dissolving in Edwardian Britain, and some farmers, at first reluctant to hasten the passing of old certainties, in due course accepted the necessity of their own self-defence, not least because agricultural trade unionism was again stirring.

The tide began to flow towards the formation of a professional association even before the First World War broke out. The NFU early discovered many functions apart from political lobbying by instituting legal and insurance services to members. But its political influence was crucial. In this respect, after an uncertain, fractious beginning which caused the Union to miss opportunities for influence, both during the War and in the depression of 1921–4, the NFU acquired a more or less coherent set of policies favouring agricultural support. By 1940 it was well placed, both through growth of membership and through the convergence of national policy with Union prescriptions, to exploit the opportunities offered by central planning to an organized interest-group. For ten years after 1943, when James Turner won a signal victory over the Ministry on the issue of an annual review of prices and subsidies, the NFU was almost an arm of the state. It entered enthusiastically and profitably into the corporatist management of agriculture. In the age of physical controls, induced modernization, and political interference, the Union assumed, and was conceded, the role of agent on behalf of all farmers. Things began to change after the end of rationing when governments repented of their previous commitment to pay for the production of food at all costs and affronted the NFU with demands for economy and circumspection. Difficult as negotiations often became after the mid 1950s, the continuing bureaucratization of agriculture and the complexity of the price-support system, before and after 1973, sustained the NFU in its duty to speak for and explain to the managers of the land.[18]

By 1950 the NFU had usurped the role of landowners' associations in negotiating for the landed interest. By the mid 1950s, indeed, most landowners were members of the NFU as farmers in their own right. As the burdens upon *rentier* landlordship piled up in the age of planning more and more of the surviving estates were taken in hand wholly or in part. Within a generation the distinction between old money and new in agriculture had become blurred; farmers as freeholders saw no great difference between themselves and neighbouring tenants and this rough equality of status gradually affected the pretensions of the old gentry as well. The new *rentiers* of the post-war world, usually business enterprises, were not disposed to practise *noblesse oblige*, so that the old relationship between caring landlord and deferential tenant fractured. Organizations such as the Country Landowners' Association, fearing for their future, actively canvassed among the lesser owners of rural property, trawling in the same waters as the NFU. Neither association became demotic; the leaders of both were still customarily men of property, education, and social weight presiding over a more or less common stock of members.

Other initiatives drew practical farmers into positions of official responsibility. The wars were especially educative. The War Agricultural Committees, though not specifically designed to be managed by farmers, usually fell into the hands of those who knew about progressive agriculture in the field. In addition, various commodity committees were dominated by the food producers. Even outside the realm of the command economy there were opportunities for farmer-directed enterprises to take root. The Marketing Boards are the best examples of producer-controlled institutions, even though they were less firmly under the control of the agricultural interest after 1954 than their predecessors had been in the 1930s. Ironically, it was the often stormy public debates about the utility or efficiency of co-operative marketing that gave those who opposed interventionism the best opportunity to make themselves heard.

Because farmers, or at least those appointed to speak for them, have been drawn so much into public affairs since 1916, their articulacy and the range of their knowledge have become issues of widespread concern. Publicity and the dissemination of periodicals and official circulars indicate both the extent of the problem and the degree of success in resolving it. By the 1950s few farmhouses were not in receipt of the NFU's own journal or one or more of the commercial magazines; and a constant stream of pamphlets – advisory, minatory, or improving – flowed from the Ministry and its affiliates. The supposition was that farmers might read and learn from such material, which in turn was founded upon a belief in their educational attainments that would have surprised Edwardian observers.[19]

It is true that the Victorians had put into effect proposals for agricultural education and training which can be traced back to the 1760s. Opportunities for farmers in general to benefit from this initiative, however, were long delayed. Before 1914 very few agriculturists were educated beyond elementary level, and almost all those that were, had not proceeded beyond local grammar schools. Social surveys in the 1930s and 1940s still indicated that formal education after school remained rare among farmers. 'Pupillage', however, was normal, although for most young men this merely meant that they learned the state of the art at their fathers' side. A detailed analysis by the Manpower Working Group of NEDO in 1972 revealed that only about 10 per cent of farmers had experienced secondary education, and that paper qualifications, at least among the self-employed, were not highly regarded. The trends in the survey disclosed a significant difference between generations. Farmers or their sons in waiting who had experienced the post-war transformation of rural education were much more likely to possess qualifications of some kind and a proportion, still apparently below 20 per cent, had attended courses in tertiary education or training. At the same time, most farm managers under 40 had been to college or agricultural institutes as part of their work experience.

The most recent information, from the mid 1980s, confirms this trend. Learning the practical side of farming as a pupil remains popular; indeed it is a requirement of most professional courses. But theoretical or scientific training is now much more commonplace among farmers or their sons. Perhaps two-thirds of those under 40 in 1985 had received some tertiary education, and many certainly also attended conferences or called in expert advisers as part of the routine of their lives. This change partly reflects the decline in numbers of landholders, not least because the principal casualties of the new engrossment have been the family farmers and the principal beneficiaries have been men in 'agri-business'. In the late 1960s the number of individuals leaving universities, colleges, and farm institutes to enter agriculture did not exceed 2,400 per annum, while the annual recruitment of farmers, managers, and workers was about 17,000. Those qualifying have less than doubled in the past fifteen years, whereas annual recruitment into the industry has fallen by half or more, so that a notably higher proportion, especially of entrepreneurs and managers – perhaps touching 90 per cent – has benefited from extended education in recent years. It is impossible to say how many farmers there were among those attending day-release courses, nor for how many these were supplementary to formal training schemes, but at its peak in the early 1970s output from such courses exceeded 12,000 a year.[20]

I suggested at the beginning that the class of agriculturists we describe as farmers survived better into this century than most of the other groups involved in cultivating the land. That they have become centre-stage, albeit sharing the limelight with politicians and officials, underlines this observation. But no latter-day farmer entrapped in the vagaries of the Common Agricultural Policy, could imagine himself in the garb of John Hodson or Fred Vincy, not to mention William Howitt's nameless midland peasantry. The social bearings of farming are set so differently from their position even a century ago that we can link the contours through time only with poetic licence. That there is a continuity in the rhythms of husbandry is more a matter of assertion than of demonstrable truth. It consoles sentimentalists to believe that the way of life in farming follows ancient landmarks. But even the worm turns soil that owes more to artifice than to the unending toil of the husbandman.

## Notes

1 Whitby, 1966; Britton and Ingersent,1964; Gasson, 1974b; *Census of England and Wales*, 1911,X, xiv-xix; *Census of England and Wales*, Economic Activity County Leaflet, Industry and Status, 1971, table 3; *Census of Scotland*, 1971; ibid., 1981; Ministry of Agriculture, Fisheries and Food, *A Census of Agricultural Statistics: Great Britain, 1866–1966*, 1968, ch. 3; MAFF, *Agricultural Statistics*, yearly, (a) England and Wales, (b) United Kingdom. Cf. Hirsch, 1951.
2 Ashton and Cracknell, 1961; Gasson, 1974b; A. Harrison, 1975, *passim;* A. Harrison, 1965.
3 Gasson, 1966a; Gasson, 1966b; Wormell, 1978, 403–10.
4 Coppock, 1971, 83–7; Holderness, 1985, 123–5; Gasson 1969.
5 See, e.g., Holderness, 1985, ch. 7; Wormell, 1978, 125–72; Sutherland, 1968, *passim;* Hallett, 1960.
6 This material, based on analysis of estates in different parts of the country, will appear in due course in my *The English Farmer: A Social and Economic History Since the Seventeenth Century*, to be published by Manchester University Press, 1990.
7 Nalson, 1968; Gasson, 1969.
8 Holderness, 1985, 129; Wormell, 1978, pt I, especially 200–18.
9 Taken from a sample of 376 respondents to a questionnaire circulated in 1978 to farmers in Essex, Norfolk, and Leicestershire.
10 This evidence is drawn from Norfolk and Lincolnshire, and since it is specific to kin-groups of my acquaintance, it is not random.
11 Holderness, 1985, 138–42, 126–7; Wormell, 1978, 125–52, 200–18.
12 Holderness, 1985, 125–6; Wormell, 1978, 410–15; Bellerby, 1956.
13 W. M. Williams, 1956, especially chs 1, 14; W. M. Williams, 1963, pt I; Littlejohn, 1963, chs 2, 7; Frankenberg, 1957, ch. 2; Emmett, 1964, ch. 4.
14 Newby, 1979.
15 See, for example, Hartley, 1937.
16 See, for example, Body, 1982; Christian, 1966; Shoard, 1980. Also Orwin, 1945.
17 See Cooper, 1980.
18 Cooper, 1980; Self and Storing, 1962; Bowler, 1979.
19 Williams, 1960, *passim*.
20 Holderness, 1985, 131–2; NEDO Manpower Working Group, *Agricultural Manpower in England and Wales*, 1972; Williams, 1960; Wormell, 1978, 541–56; Bessell, 1972; Whitby, 1967; Giles and Mills, 1970.

# 8

# The most despised craftsmen: farmworkers in the twentieth century

W. A. Armstrong

The First World War bit deeply into the fabric of rural society and farmworkers were not insulated from its effects. An initial outflow of young men animated by patriotic fervour or the agreeable prospect of a change in their normal routine was succeeded, in the later stages of the war, by waves of conscripts. Roughly, some 36,000 farmworkers from England and Wales lost their lives, and two or three times as many returned nursing wounds of varying degrees of severity which in some cases would trouble them for years to come.[1] On the home front, the exigencies of wartime food supply heralded some remarkable if temporary shifts in the composition of the farm labour force. While the main burden was shouldered by established workers, retired men were also brought back into service. Their skills, several commentators agreed, were fully appreciated for the first time as they were relied upon to lead and instruct various classes of auxiliary labour – including large numbers of village women, the Women's Land Army, prisoners of war, and soldiers – who were often unfamiliar with farm work.[2]

The war put an end to piecework in many districts and reduced the irregularity of employment: it also heralded developments which would change the character of the agricultural industry. Firstly, these years saw the appearance of farm tractors. Mostly imported from the USA, and being initially unreliable, they were widely regarded as twenty-minute wonders, and at this time they did not make a major contribution to upholding food production; but they were the harbingers of a much more rapid rate of mechanization, based upon the internal combustion engine, which eventually brought about changes more profound than any achieved through the use of steam engines. Secondly, the war saw government intervention on an unprecedented scale. The agencies through which policy was translated into action were the County War Agricultural Committees, while central control and direction culminated in 1917 with the Corn Production Act, which guaranteed cereal prices and set up an Agricultural Wages Board to determine the level of farmworkers' pay. Under its aegis, a tendency for wages to decline in real terms on account of rapidly increasing wartime prices was arrested and regional differences were reduced. It is true that attempts were made to put this policy into reverse in the straitened circumstances of the early post-war period. Similarly, the War Agricultural Committees did not, of course, survive, and in 1921 a second Corn Production Act (that of 1920) was revoked along with the statutory determination of wages. However, this return to laissez-faire economics did not last for long. Precedents had been set and a series of *ad hoc* steps were taken during the inter-war years to uphold prices and wages in an industry which could fairly claim to be depressed, while during and after the Second World War agriculture came to enjoy undreamed of levels of state support.

An examination of the size and composition of the labour force offers a suitable starting point for further analysis. After 1918, there is evidence to show that many demobilized servicemen were reluctant to return to the land. Even so, agriculture remained a major employer; the 1921 census indicated a fall of only 8.8 per cent since 1911, and a total of 869,000 employees (8.4 per cent of these being women and another 21.2 per cent classed as casuals) still earned their living from the soil.[3] The age composition already noted as characteristic of agriculture (see pp. 38–9) was still lop-sided; there were armies of elderly workers, and youths were not difficult to come by. Retaining them was another matter. A scrutiny of the destinies of leavers from eleven Warwickshire schools in 1918–27 revealed that only 36 per cent of the boys were engaged on farms at the later date, even though the villages concerned appeared to offer virtually no alternatives. Likewise, a small study of 36 boys who left school in the Cotswolds at the age of 14 in 1921, and whose whereabouts were known a decade later, is illuminating. Nine had left their native villages; these included 2 soldiers, 2 policemen in Birmingham, a stonemason, a bootmaker, a jobbing gardener, and 2 agricultural workers. Of the remaining 27, 9 were farmworkers, but the occupations of the rest varied: they included 2 estate gardeners, 2 gamekeepers, a groom, a footman, 2 carriers, a grocer, 3 handymen, a lorry driver, and a garage mechanic.[4]

On a larger canvas, it could be inferred from comparing the censuses of 1921 and 1931 that something like one-fifth of 276,000 males aged 15–24 at the earlier date had been lost to agriculture.[5] Nevertheless, the reduction in the total farm labour force in the 1920s was quite small: the economies sought by farmers as prices plunged appear to have impinged chiefly upon casual employees, both male and female, although large numbers of seasonal workers continued to be drawn from London and other large towns to assist with the fruit, hop, and potato harvests. The 1930s were to present a somewhat different picture. The number of full-time male workers plummeted sharply, decreasing by 17 per cent in only nine years.[6] Significantly, the loss was least apparent in counties characterized by coal-mining and heavy industry, and in Wales where there were even signs of a return to agriculture by some such workers. It was in the southern counties, hitherto relatively untouched by industrialization, that the decline was most obvious. Cases in point included Oxfordshire, featuring the Morris works at Cowley; Berkshire, where the MG plant at Abingdon was a special attraction; and Buckinghamshire, where the outflow from agriculture was ascribed to the movement of labour into furniture factories at High Wycombe, paper mills near Woburn and Bourne, and the brickworks in the north-east of the county.[7]

The Second World War arrested, indeed reversed this tendency. In a scenario that in many respects repeated the experience of 1914–18, the nation was obliged to give high priority to agricultural production. Farmworkers were subjected to a variety of controls upon their movements and, in general, were not taken for military service unless they volunteered. Once again it fell to them to direct armies of auxiliaries, notably land-girls (the Land Army was re-established in 1939), soldiers, and after 1942, prisoners of war. The last prisoners were not repatriated until 1948 and the Land Army remained in existence until 1950; yet it is striking that in 1949 the number of regular full-time male employees (595,000) was higher than in any year since 1933.[8]

However, it was not long before the pre-war trend began to reassert itself. Through the years 1949–65, the rate of decline was about 4 per cent per annum, and in just thirty years, 1949–79, the number of full-time agricultural employees fell by no less than two-thirds. There have been numerous studies of this outflow; J. D. Hughes, for example, shows that the rate of employee retention was low in agriculture compared to other industries, particularly for men under 30; while another study estimates that no fewer than 43 per cent of those under 25 in 1953 had been lost to British agriculture ten years later, on a net basis.[9] The upshot of these changes was that by 1971, even in the rural areas of Norfolk (i.e. the administrative county, excluding its boroughs), male agricultural workers did not account for more than one job in nine; and in some rural districts in the county, for as few as one job in thirteen (Blofield) or one in nineteen (St Faiths).[10] On the other hand, the recent increase in the owner-occupation of British farms (see pp. 104–5) has been paralleled by a mounting reliance on family labour. In Staffordshire hired men were thought to account for only 30 per cent of all men working on farms in 1973, and by this date Bessell was expecting family labour to become more and more significant. It is impossible to trace the shift in detail, but one recent estimate suggests that if we add to the labour of relatives the contribution of farmers and their wives, the proportion of 'standard man-days' provided from within the family already exceeded 40 per cent in the 1960s, and reached over half in the ensuing decade.[11]

An immediate explanation of the huge reduction in the hired labour force is mechanization, which had not progressed during the inter-war years at the rate once considered likely, but which advanced after the war at a heady pace. Tractors, the epitome of versatility, first surpassed horses in 1950, and rose in numbers to reach 435,000 by 1965, after which they stabilized in numbers if not in size. Other machinery, such as combine harvesters and milking parlours, multiplied rapidly, eventually lifting the amount of capital per man well above industrial norms, but implying substantial reductions in the quantum of labour required. However, comparatively few men were 'driven' from the land. Much of the labour dispensed with was either female or part-time, and reductions in the full-time male labour force were often achieved through natural wastage, as workers retired and were not replaced.[12] Moreover, prior to the 1970s, a state of more or less full employment in the economy at large made it easier for school-leavers to find alternative jobs; only 4.6 per cent entered the industry in 1970, against 9.2 per cent in 1950.[13] In addition, young men with a few years of experience in mechanized agriculture found that their skills, such as driving, were taking on an increasingly transferable character. Of course, the extent to which alternatives were available varied significantly from one district to another, and over time. In the late 1960s, school-leavers in remote Norfolk villages were still liable to be locked into agriculture until they were of an age to join the armed forces or secure a driving licence, and speaking of Suffolk in the 1970s, Newby remarks that ascription was by no means a thing of the past.[14] Elsewhere, for example around Birmingham, Coventry, and Scunthorpe in the 1950s and 1960s, and also in the vicinity of new towns such as Stevenage and Milton Keynes, more opportunities presented themselves. However, this did not mean that farmers in these neighbourhoods were starved of labour; they might well draw mature men from further afield, attracted in part by non-farming jobs available for their adolescent children.[15]

In general, without denying the possibility of cases of individual hardship, the reduction in the size of the hired labour force has been achieved relatively smoothly. The influence of mechanization is particularly difficult to assess. Some of the evidence discussed above, taken together with the farmworker's manifest disadvantages, suggests that it should be seen as the facilitator rather than the direct cause of declining numbers. However, machinery has certainly effected many important changes in the nature of farm work and it is this aspect to which we now turn.

Already by 1924 the Liberal Land Committee concluded that farm work was becoming 'less skilled', and a perspicacious correspondent of Robertson Scott noticed a tendency for job specialization to die away, along with the ultra-conservative attitudes that had accompanied traditional skills.[16] Thus, Fred Kitchen remembered how all lads of the pre–1914 era in south Yorkshire had been 'horse-proud', but remarked on the disappearance of this attitude by the 1930s.[17] The years following the Second World War saw increasing numbers designated as 'tractormen' and the most obvious manifestation of technological change was the compression of harvesting, which in most fields became unlikely to take more than a day with the aid of the combine, the grain being emptied into lorries and carried away in bulk, the straw and stubble being promptly baled and removed, or even burned. Tasks such as hedging and ditching were rendered less necessary, or let to contractors; spraying cut down on singling and weeding; and dairy cows were milked mechanically using assembly-line methods.

In many respects the division of labour has shown a marked tendency to become more rudimentary. Farm work has also become more lonely and isolated. 'Out in the fields the lone tractorman can be seen ploughing, drilling, spraying, baling or whatever the season requires', says the historian of Corby Glen (Lincolnshire), going on to mention the rarity of team-work in any form.[18] Along with these changes, entailing the demise of the horse, has gone a dramatic decline in orally-transmitted traditions and culture, which George Ewart Evans began to record, in the nick of time, in the 1950s.[19] All this is true, but we need to be wary of equating these developments with declining skills, still less with the notion of the descent of farm labour to a dead level of mediocrity: as Lennard observed presciently before 1914, one should not call Moltke a degenerate warrior on account of his inability to throw a boomerang.[20] At any given point in time the old skills and the new tended to be the perquisites of different generations and young men were often observed to take pride in their machinery, seeing it as a badge of modernity. The skills involved in manipulating agricultural machinery efficiently and safely can easily be under-estimated, as is the range of demands made upon the modern farmworker. In 1951 J. R. Bellerby addressed a questionnaire to men who had worked in both agriculture and industry; from their responses to questions about the skills, responsibilities, and stresses involved, he concluded that farm work was at least as demanding as in the average industry.[21]

In these circumstances it might have been expected that more formal or specific courses of training would emerge, along with a suitably graded career structure to replace the vanishing hierarchies characteristic of times past. However, progress was disappointingly slow. By 1938 degree or diploma courses in agriculture were on offer at nine British universities and forty-three Farm Institutes, while here and there, as at Chadacre in Suffolk, farm schools offered short courses of a less ambitious nature. But there were few scholarships and the subject excited little public interest.

The upshot was that only about one in every twelve farmers of the late 1930s had received any formal training, and the number of workers who had enjoyed anything of the kind was infinitesimal, complained Pedley, contrasting the position with that in Denmark.[22] Even after the war when it had become quite the fashion to proclaim from public platforms that brains rather than brawn were requisite in the new agriculture, progress still remained slow, not least because farmworkers were well aware of the low value set on qualifications by employers. Thus, an apprenticeship scheme started in 1953 succeeded in enrolling only about 200 youths per annum, and only about 4 per cent of the 2,359 students registered for one-year courses in the mid 1950s were from farmworkers' families. In 1966 it was noticed that only 8 per cent of new entrants into the industry in Yorkshire were registered apprentices, and employers still evinced a strong tendency to say that what distinguished one worker from another was not so much training as 'character'.[23] For years, discussions aimed at producing a progressive career structure were bedevilled by this attitude, well rooted in the National Farmers' Union; also by the suspicion, in the workers' unions, that any such arrangement would serve only to set man against man. However, by the early 1970s some progress towards a simple career structure had been made. The scheme adopted in 1971 featured four adult gradings: I and II (these were supervisory categories), craftsman, and ordinary worker. The first three of these commanded wage premiums set by the Agricultural Wages Board at 30, 20, and 10 per cent respectively, and existing workers were graded according to their employers' declarations of competence. Concurrently, a New Entrant (Apprenticeship) Training Scheme was started which entailed a three-year period of practical training on the farm, integrated with part-time further education and leading to the craftsman's certificate. The ensuing years saw a marginal increase in the percentage of men classed as craftsmen, although in 1980 the majority remained in the ordinary grade.[24]

Taken in conjunction with the contraction of the farm labour force (see pp. 116–17) and the declining opportunities of social mobility (see pp. 128–30), the indications are that farm employment has continued to suffer from the 'want of outlook' remarked on by Acland in 1913.[25] Even so, when asked their opinions farmworkers usually profess a degree of satisfaction with their work. Among a group of Suffolk employees in 1972, 93 per cent considered their jobs to be either interesting all the time, or mostly interesting; and among young men interviewed for an official enquiry of 1970 many said that they had entered the industry because they were 'always interested in farming' and only 1 per cent actively disliked the work.[26] In the farm context, it seems that mechanization does not have the same deadening effect as on the factory assembly line, if only because the operator is the director rather than the servant of the machine. In addition, nature dictates that farming operations remain sequential: the seasonal rhythms of work still have variety and meaning, and the inherent satisfaction of seeing animals and plants grow is not to be discounted.

Moreover, those who have elected to remain on the land can usually exercise a degree of choice as to their employers; the notion that workers remain on the same farm throughout their lifetimes is not true for the majority. Only very fragmentary information is available for earlier periods, but it seems probable that the Second World War marked an acceleration in the rate of turnover. Between 1940 and 1943 at least 38.3 per cent of employees on a group of southern farms changed employers at least once, and in Buckinghamshire in 1946–7 the figure (on 179 holdings) was over 25 per cent.[27] At Elmdon in Essex, where in 1930 there remained numerous farmworkers' families of at least sixty-nine years standing in the village, jobs changed hands much more frequently by the 1960s and farmers had to look far afield for new men.[28] Although, as we have seen, many departed from the industry, some 40–50 per cent of workers leaving one agricultural situation took up another in their search for marginal improvements in wages, promotion possibilities, a better standard of accommodation, or places situated near towns; and most such moves were initiated by the workers themselves.[29] Such figures dispel the notion that the farmworker has no freedom of movement. On the other hand, rates of turnover are not especially brisk by comparison with other industries, such as engineering in Birmingham, and some evidence of the close ties of employers and some employees is afforded by Newby's information gleaned from Suffolk farmworkers in 1972: he found that the average length of service was 16.7 years, while at the national level it appeared that 62 per cent of employees, undifferentiated by age, had served five years or more on the same farm.[30]

During the inter-war years the labour force fell by more than did farm output, and productivity is believed to have increased by some 40 per cent between 1924 and the outbreak of the Second World War. Thereafter, the amount of capital supporting each agriculturist soared and, since about 1960, advances in labour productivity have averaged about 6 per cent per annum, against less than 3 per cent in the British economy as a whole.[31] These circumstances facilitated gains in wages and living standards, at a rate which was unprecedented; however they were not automatic in operation, nor were the results so satisfactory as many farmworkers would have wished.

The loss of the Agricultural Wages Board of 1917–21 was a blow to labour, and the voluntary county conciliation committees which replaced it turned out to be ineffective. Wages fell disastrously as farmers struggled with depression prices, from 37s. (ordinary workers) in December 1921 to 28s. a year later, and they remained at that level through 1923 and

1924. The critical situation facing the agricultural industry was, however, the subject of an enquiry appointed by Bonar Law's government in 1922, and it was against this background, with the Norfolk strike (see p. 128) and repeated demands from farmworkers' organizations, that the first Labour government brought forward in 1924 a bill to restore statutory wage legislation in agriculture – though not the guaranteed prices with which this had been inseparably linked in 1917–21. The new arrangement placed particular responsibility on county committees (each consisting of equal numbers of representatives of the NFU and of Labour), which were given powers to determine wage rates, hours, and conditions, as well as to grant permits of exemption for non-able-bodied workers.

The new measures were not considered an unqualified success; test inspections revealed that as late as 1935–6, the proportion of workers who were paid less than the agreed minimum was as high as 21 per cent in England and 36 per cent in Wales.[32] Even so, wages rose a little in 1925 (by some 3s. 5d. to reach 31s. 5d.) and thereafter remained remarkably steady. Indeed, after 1934 they began to ascend slowly, reaching 34s. 9d. in 1939, while the tendency for regional contrasts to diminish continued, so that in 1937 minimum wage levels were no higher in, say, Durham and Northumberland than in Herefordshire or Wiltshire.[33] Many farmworkers earned significantly more than the wages just quoted, if only by virtue of the extra hours they were obliged to put in, and this factor, together with the reduced cost of living (in 1925 it stood at 175, and in 1938 156 on the base 1914 = 100), ushered in a modestly wider range of consumption. Among the novel satisfactions found in farmworkers' homes at this time were oil stoves, mantle lamps, more utensils, wireless sets, bicycles, and motor cycles, the last being the prerogative – usually – of the young and single.[34] However, it was unlikely that any farmworker's family could enjoy all these simultaneously, and incipient consumerism was braked by shortcomings in basic amenities such as electricity, by difficulties of access to shops, and in many cases by sizeable families. Food still loomed large in most household budgets, accounting for no less than 63 per cent of all disbursements by farmworkers in Lincolnshire during the spring of 1937, which was actually higher than a comparable figure calculated for the unemployed in the Rhondda Valley (49 per cent). An extensive official survey of household budgets of 'rural' and 'industrial' workers taken in the same year revealed that although the former spent less in real terms on medical attention, cleaning materials, smoking, and newspapers, it was at least as high a proportion of their more limited incomes; while they laid out less both in real terms and as a proportion of their earnings for footwear, clothing, travel, and entertainment.[35]

The Second World War necessitated significant changes in the operation of measures concerning farm wages. In order to stabilize the supply of labour, an Agricultural Wages (Amendment) Bill of 1940 empowered the

Central Board, which had hitherto acted as a kind of cypher, to fix a national minimum wage which was set at 48s. in May of that year. This substantial increase was secured only by compensating farmers to the tune of a £15m addition to authorized prices, and through the rest of the war wages continued to be adjusted so as to correspond almost exactly with farm prices. By 1945–6 the minimum wage had reached 72s., and the total earnings of many farmworkers stood a good deal higher than this: in East Anglia in 1945 tractormen earned 77s. 7d., horsemen 84s. 2d., stockmen 87s. 8d., and cowmen 93s. 10d., although in every case they would have been putting in hours far longer than those specified for the minimum wage.[36] In Kent, so it was alleged, some men in 1944 told their employers that if they had further increases they simply would not know what to do with the money, an attitude firmly deprecated in the local trade union organizer's monthly newsletter.[37] The war years are remarkable for the fact that, in contrast to the late nineteenth century and the inter-war years, real wage gains were made at a time of rising prices; but as leaders of the agricultural unions were doubtless aware, farmers fared better. Their net incomes were particularly buoyant, exceeding the increases for wages in general, salaries, professional earnings, and company profits.[38]

As reconstructed in 1947, the Wages Board comprised eight representatives from the NFU, eight workers' representatives, and five independent members appointed by the Minister of Agriculture who in practice acted as arbitrators, and this system of agricultural wage determination has prevailed ever since. Over the years, the gap between minimum wage levels and actual earnings has tended to widen, and the average farmworker could expect to receive £5 9s. in 1948–9, and thirty years later, in 1979, £72.04, with a quarter earning less than £53.34 and a quarter over £82.53.[39] After account is taken of cost of living increases it is possible to infer a 27 per cent gain between 1948–9 and 1960–1, with further improvements of 25 and 26 per cent for the 1960s and 1970s; so approximately doubling farmworkers' real incomes since the war.[40] By the late 1940s a greater variety of household goods, notably electrical appliances such as irons and fires, were beginning to trickle into farmworkers' homes and, with only a short time-lag, cottages began to boast television sets. In 1961, 80 per cent of estate workers at Lockinge (Berkshire) owned one, and by this date 37 per cent also had cars, which had been exceptional there only ten years before.[41] It was a sign of the times when in 1966 the Wetherby branch of the National Union of Agricultural Workers drew attention to the ample parking available at the venue of its annual dinner and dance, while in Kent in 1972, members were offered stickers for their cars passing in cavalcade through Maidstone, Canterbury, and Ashford in support of the current wage claim.[42]

The advances observed were considerable, but judged on a comparative basis the story is a rather different one. Towards the close of the inter-war period the average wage (for ordinary workers) was 33s. 7½d. (minimum) and 34s. 3d. (total earnings), which compared poorly with labourers employed in building, shipyards, on the railways, in engineering, or by local authorities (47s. 0d. to 54s. 4d.), and even more unfavourably with those of skilled workers, such as bricklayers, carpenters, and masons (71s.), printing workers (73–4s.), or railway engine drivers (72–90s.).[43] Matters improved somewhat during the Second World War and its immediate aftermath, since between 1938 and 1949 minimum wages in agriculture increased by 170 per cent, against 81 per cent for the average of all industries.[44] However, this was not destined to last, and the farmworkers' earnings soon showed a renewed tendency to fall behind. According to one study, between 1947–8 and 1955–6 workers in 'all industries' saw a gain in the real value of their earnings which was twice as great as that of farm employees, i.e. 18 per cent against 9.5; while a longer-run calculation suggests that through the period 1949–72 average weekly earnings increased by 291 per cent in agriculture and 376 per cent in other industries.

This contrast is exacerbated if account is taken of hours, so that relative to industrial wages the hourly earnings ratio slipped back from 69 per cent in 1949 to 60 in 1964, where it still hovered in 1972.[45] At about this time Bessell noted that farmworkers' earnings would need to be increased by about a quarter to give them parity with lorry and van drivers; and on an hourly basis the incomes of farmworkers were exceeded by 29 per cent (in the case of lorry drivers), 53 per cent (postmen), and 93 per cent (policemen).[46] Although subsequently there were signs that the new wages structure was beginning to affect the relative earnings of farmworkers,[47] the shift was not large. They were accordingly the subject of attention from the Low Pay Unit, which in 1975 pointed out that a quarter of families with children drew Family Income Supplement, while fewer than half had enjoyed a holiday away from home the previous year, and many relied for clothing on second-hand sources of supply and jumble sales. The report also pointed out that motor cars had become a practical necessity; and that according to their survey, only a quarter of wives were able to work in order to supplement the family income. This unusually low proportion was due, in the main, to a paucity of opportunities and poor public transport facilities.[48] Thus, for all that conditions had undoubtedly improved, the disparities with the living standards of farmers, urban workers, and perhaps most visibly, neighbours who worked outside agriculture remained entrenched.

The style and standard of living of farmworkers has been influenced by a variety of factors other than wage levels, and some of these are discussed in other chapters. Here it is possible to consider only three of them, namely, health, housing, and education.

During the nineteenth century the longevity of farmworkers was comparatively good. In 1930–2 age-standardized mortality rates among all males engaged in agriculture were still well below the national average; indeed, they were a shade lower among farmworkers than among their employers, who showed a greater proclivity to commit suicide and were more prone to succumb to diseases associated with good living. As late as 1970–2 death rates among farmworkers were about 10 per cent lower than those expected from males of similar social class.[49] Nevertheless, certain health complaints were linked with the nature of the work (notably rheumatism, arising from repeated exposure to cold and wet conditions), and in the 1930s, agriculture exhibited quite a formidable list of health risks, although admittedly, ones which workers shared with their employers: these included an above-average propensity to tuberculosis, associated with the ingestion of untreated milk, and sundry diseases communicable from animals to men, such as anthrax and ring-worm. There were also toxicological risks linked to the application of scientific products to farm work, including mercurial products used in treating seeds, arsenical compounds used as sheep-dip, and certain insecticides;[50] while in more recent years, the increasing chemification of agriculture has been received with some suspicion, the NUAW's pamphlet *Pray before you Spray* (1981) strongly advocating a complete overhaul of the system of pesticide control in Britain. Nor should the accident rate be overlooked. According to one calculation, farmworkers in 1947 stood a one in twenty chance of sustaining injuries that would keep them from work for seven days or more, ladder work being a prime cause, and although the mechanization of agriculture was eventually accompanied by numerous regulations (notably for tractor cabs in 1967 and 1976), in 1980 as many as 4,248 employees were reported injured on British farms, including twenty-four fatalities.[51]

These points suggest that agriculture is by no means so free of hazard as is popularly imagined and, so far as general health and longevity is concerned, the farmworker's once significant comparative advantage has almost certainly dwindled. Primarily, this reflects the fact that during the present century so many of the basic deficiencies of nineteenth-century towns have come to be tackled more effectively, one of the most obvious signs of this tendency being a rapid improvement in the general standard of British housing.[52] Even in the rural areas, some progress was made, for 266,000 new houses were constructed with state assistance between 1919 and 1938. Meanwhile, under a series of Acts commencing in 1926, landlords were assisted by grants or loans to improve or recondition some 13,783 cottages by 1937, and there was also some grant-aided demolition of houses judged to be particularly bad.[53] The results of such activity were sometimes aesthetically disturbing. Thatched roofs might be replaced with Welsh tiles or, worse, corrugated iron sheets, and in one North Oxfordshire village an investigating group remarked on the disappearance of a whole street of seventeenth-century cottages of solid construction and beautiful design; their inhabitants had been rehoused in a council colony at the edge of the village, 'designed without the smallest consideration for the traditional architectural types of the locality'.[54]

Doubtless the average quality, if not the rustic appeal, of rural housing was enhanced during these years, but it would be erroneous to suppose that farmworkers were among the chief beneficiaries. According to W. H. Pedley, only about one new house in fourteen was occupied by a farm-worker, for they found it difficult to meet even council-house rents on their limited wages; and in Devon, whose record in reconditioning was the best in the country, not more than half the improved houses were so occupied.[55] By and large this segment of the rural population continued to endure the disamenities inherited from the past, for when all the improvements are allowed for it remains true that in 1939 25 per cent of all parishes – to say nothing of isolated hamlets and cottages – lacked a piped water supply or a sewerage system.[56]

The Second World War, like its predecessor, marked a pause in housing improvements, but thereafter they came at a rate which was brisk by historical standards. The usual sequence was, first, the provision of cold water taps, second, electricity, and third, the introduction of a water-borne sewerage system to replace the formerly ubiquitous earth-closets or privies, whose utility (as a source of garden manure) had long been obvious only to those who rarely, if ever, had to use them. Thus, the average standard of amenities in farm cottages was gradually brought up to a level approaching that of local authority housing. However, the issue of quality was not the only problem. What has loomed large in the eyes of critics is the basis on which tenancies are frequently held, namely, the tied cottage system. This arrangement has a long history and the restrictions it could impose on a worker's freedom of action ensured that it would be opposed by the farmworkers' unions, even before the First World War. Paradoxically, although employers in the present century have been less inclined to take advantage of tied cottages as an instrument of labour discipline, and evictions have become comparatively rare, the system has been regarded increasingly as a 'feudal' relic. Generally speaking, farmworkers have not enjoyed incomes at a level enabling them to join the rush to owner-occupation that has been so marked a feature of recent decades, and although the number living in tied cottages has decreased substantially, the proportion doing so has increased; from about 34 per cent in 1948 to just over 60 per cent in the early 1970s, when fewer than 10 per cent of farmworkers owned their own houses, 11 per cent lived in other rented accommodation, and the rest resided chiefly in council houses.[57] Although surveys showed that resentment of the system was very far from universal,[58] it may well have acted as one of the factors prompting some to leave agriculture; and, despite the introduction of various restrictions on the farmer's freedom of action by Acts of 1964 and 1970, not a few farmworkers, or their surviving dependants, have had reason to regret their situation in later life.

Another area of relative disadvantage was in basic education. Although by the inter-war years changes in the rural occupational composition ensured that the majority of pupils in most village schools were not farmworkers' children, virtually all farmworkers' children began and ended their education in village schools. Frequently these were 'all-age' institutions; that is, they retained their pupils until leaving age (normally 14), for the sequence of 'junior' and 'senior' education recommended in the Hadow Report of 1926 came closer to realization only in the towns during the 1930s. Village schools were criticized for a variety of reasons. Their teachers, though frequently commended for their devotion, tended to be less well qualified and were accordingly lower paid than urban teachers; classes were frequently housed in antiquated buildings with inadequate facilities; cultural – and, to an extent, material – poverty ensured that the 'apathy' of many parents was passed on to their offspring.[59] Moreover, there was a certain confusion of purpose at the heart of village education: the issue of whether these schools should have a 'rural bias' to the curriculum had never been satisfactorily resolved, some arguing that teachers should be 'apostles of agriculture' and others considering it essential to convey the idea that the world was larger than the farm.[60] Yet, for all their defects village schools did succeed in some cases in laying a basis for reading, as appears to be evidenced by the rising demands upon the library services which the county authorities were empowered to support from the rates after 1919;[61] and although country children tended on average to exhibit lower 'intelligence quotients' than their urban counterparts, an increasing number of critics of such 'scientific' tests were aware that they could not be made culture and value free, and in particular warned against confusing a patient, watchful stolidity of manner with dimwittedness.[62]

The Second World War occasioned an intended postponement of the raising of the leaving age to 15 and brought numerous shortages of teachers and classroom materials, these problems being exacerbated in many cases by the arrival of urban evacuees. Thereafter, change was exceedingly rapid. In the main, small village schools became 'primary schools' and, in implementing the Butler Education Act of 1944, children were bussed out of villages on reaching the age of 11 to pursue their 'secondary' education, some in grammar schools but the majority in secondary modern schools prior to the appearance of comprehensive institutions from the 1960s. 'Miss Read''s *Village School* (published in 1955) gave an authentic picture of the facilities and atmosphere of small rural schools at that time and, commented a reviewer, afforded 'no inconsiderable argument for the retention and improvement of our village schools'.[63] This sentiment has not, however, prevented many disparaging attacks on them, even in their reduced role as feeder schools, by educationists; and on a variety of grounds, not least to achieve economies, thousands of village schools have been closed since the 1950s. As a consequence of these various changes, the offspring of farmworkers have certainly enjoyed a longer period of education than their forbears and one which, in principle, has offered more opportunities. The minority of farmworkers' children who secured good educational qualifications in virtually all cases took up careers in other fields, but even the majority of children who were not outstanding achievers in these terms were exposed to a wider range of influences, inducted into the habit of commuting from an early age and, as the dwindling number of farmworkers shows, were in most districts far less likely to be locked into agriculture than their forefathers had been.

The twentieth-century history of trade unionism in agriculture offers a useful insight into the self-perceptions of farmworkers. The circumstances of the First World War and its immediate aftermath strengthened the position of the National Union of Agricultural Workers, which in 1920 had attained an estimated membership of 93,000; together with the agricultural membership of the Workers' Union, this suggested that up to half the male labour force had been organized. However, the ensuing difficulties of agriculture were unpropitious, and by 1929 membership of the NUAW had dwindled to 23,000 with a mere 5,000 farmworkers in the Workers' Union.[64] Despite the rhetoric of certain left-wing leaders, such as R. W. Walker, General Secretary of the NUAW until 1928, the weaknesses of the movement were amply illustrated during the Norfolk Strike of 1923.[65] This defensive action was prompted by successive wage reductions during the short period when no statutory regulations were in force, and the farmworkers' case excited considerable public sympathy. It featured the use of blackleg labour, intimidation on both sides, and some victimization after a settlement had been patched up. Yet it is striking that even in this most strongly unionized of counties, no more than about 5–7,000 men were willing to down tools, among a labour force of not less than 30,000.[66] Moreover, although the strike was certainly a factor encouraging the 1924 Labour Government to reinstate wage-fixing regulations, it had an enduring impact on the strategies of the NUAW, by cutting the ground from under the feet of militant leaders. Under the direction of Edwin Gooch (President from 1928 until his death in 1964) a much more cautious, not to say conciliatory, stance was adopted; for example the union advocated the continuation of sugar beet subsidies in 1934 and the support of home cereal prices in 1935, both these positions being out of line with official Labour Party policy.[67] However, efforts to increase membership were not very successful, and according to Pedley only about one farm-worker in twelve was in a union in about 1939.[68]

The Second World War, like its predecessor, was conducive to growth; the NUAW tripled in size by 1945 and attained its all-time peak in 1948 (137,000).[69] Moreover, the *rapprochement* with employers, cautiously pursued before the war, was taken further. In speeches on wartime platforms to such august bodies as The Farmers' Club, Gooch stressed the identity of interest of employer and employee, characterized as 'the most despised of our craftsmen'. Anticipating that public interest would fade after the cessation of hostilities, he urged the need for landlords, farmers, and workers to stand together in defence of their industry and livelihood.[70] In the context of post-war state support for agriculture (see pp. 111–12), the position of the farm-worker visibly advanced, as we have seen, and to that extent Gooch's strategies were vindicated. But the rate of progress seemed slow, and invariably wage increases fell somewhat below union demands. It has been shown that short term fluctuations in union membership were sensitive to the unions' degree of success in this respect,[71] and in the longer term it showed some paradoxical features. Density of membership showed a tendency to increase; in Staffordshire by the mid 1920s, 60–5 per cent of farmworkers were members and the same, roughly, was true of Kent.[72] But the agricultural labour force was declining at such a rate that the absolute size of the NUAW was set on a declining trend, despite its frequent recruiting campaigns. Meanwhile, the fact that wages in agriculture tended to lag behind not only farmers' incomes but also those of other wage-earners led eventually to some questioning of the tactics of the union establishment by a new generation of radicals. However, the membership as a whole seemed indifferent to calls for militant action, and it is fair to say that the critical issue of whether the true interest of the farmworker lay alongside that of his employer, or whether he should align himself more closely with workers in other industries, remained unresolved as late as 1982. In that year, in the face of mounting financial problems, the NUAW decided to join with the Transport and General Workers' Union: about 43 per cent of members voted in favour, 7 per cent against, and half did not vote at all.[73] It is too early to say whether the merger will improve matters for the farmworker, but it is clear that the demise of a separate union is highly symbolic, marking the end of an era.

This brief recitation of the role of trade unionism in agriculture leads finally into the broader issue of the farmworker's conception of his place in society. In the 1930s the proportion of all agriculturists who were wage-earners was 66 per cent, and stood higher than in almost every other country in the world.[74] Moreover, the chances of a wage-earner moving up into the ranks of the farmers were slim, and in the post-war world deteriorated to virtually zero.[75] Viewed objectively, especially by those oblivious of variations in skill, responsibility, and status, there has been no difficulty in identifying farmworkers as a distinctive class. How far they have regarded themselves in this way is a debatable point. Certainly, the sense of being socially disadvantaged is one that has been widely shared among farmworkers during this century and the last; and levels of self-respect cannot have been improved by their constant portrayal in the media as quaint, old-fashioned buffoons. But class awareness, in this limited sense, is a far cry from class consciousness in its developed form, entailing the recognition on the part of a class that its interests are not only different from, but opposed to, those of its employers.

Even in the 1920s, when union leadership was at its most militant, most farmworkers were immune to ideological influences.[76] Later, in the 1970s, the sociologist Howard Newby encountered a mass of ambivalent attitudes when he investigated the images of society held among Suffolk farmworkers: the number of 'classes' they distinguished ranged from nil to five and there were twenty-six distinguishable types of nomenclature, though with some heaping on 'a dichotomous ascriptive model'. Far from seeing their employers as class enemies, workers had increasingly (in Newby's view) come to identify their interests with those of their bosses. Sixty per cent of those surveyed agreed that 'most employers had the welfare of their employees at heart', and any animosities expressed tended to be aimed at newcomers flocking into the villages, or, more significantly, towards workers in other industries. No fewer than 92 per cent felt that they had more in common with farmers than with non-agricultural workers, and 86 per cent reckoned that industrial employees were getting more than their fair share of wages.[77]

Findings such as these indicate the strong survival powers of the venerable concept of an 'agricultural interest' which was part of the common vocabulary of the eighteenth century and, if somewhat threatened during the nineteenth, has staged something of a revival under conditions whereby the state has come to control the fortunes of agriculture. They suggest also, perhaps, that those who are most disposed to be discontented with their life-styles and who derive little satisfaction from the work have been prominent among the many thousands who have left the industry since the war. But there is no satisfactory evidence to confirm the view, once commonly aired, that it is invariably the brightest and best who have departed, leaving behind them a residuum comprising the dull and unenterprising.

## Notes

1 For the basis of these estimates, see Armstrong, 1988, 171, 281.
2 A detailed analysis of wartime farm labour supply is provided by Dewey, 1975 and Dewey, 1979.

3 Taylor, 1955, 38–40.
4 Ashby, 1933, 231; Abrams, 1932, 64–5, 68.
5 Britton and Smith, 1947, 206.
6 See table 8.1 in Armstrong, 1988, 175.
7 ibid., 176–8.
8 Ministry of Agriculture, Fisheries and Food, 1968, 62.
9 Hughes, 1957, 38; Ministry of Agriculture, Fisheries and Food, 1967, 13–14.
10 Calculations based on data given in Office of Population Censuses and Surveys, 1975, 5.
11 *Victoria County History, Staffordshire, VI*, 1979, 144; Bessell, 1972, 3; Holderness, 1985, 135.
12 See Gasson, 1974a, 42–4, for a more comprehensive discussion of 'push' and 'pull' factors, which may have varied regionally.
13 Gasson, 1974b, 121.
14 Gasson, 1974a, 48; Newby, 1977, 195.
15 Gasson, 1974a, 25, 79.
16 Liberal Land Committee, 1925, 73; Robertson Scott, 1926, 255.
17 Kitchen, 1930, 45, 221.
18 Steel, 1979, 4.
19 Evans, 1956 and Evans, 1970.
20 Lennard, 1914, 42.
21 Bellerby, 1952, 13.
22 Pedley, 1942, 119–21; for Chadacre, see Seward, 1937, 37–41.
23 Giles and Cowie, 1964, 14, 16; Black, 1968, 69.
24 Lund, Morris, Temple, and Watson, 1982, Appendix IV, 43.
25 See chapter 4, p. 63.
26 Newby, 1977, 289–90; Bessell, 1972, 20.
27 L. G. Bennett, 1943, 86; Mollett, 1949, 188.
28 Robin, 1980, 89, 105.
29 Gasson, 1974a, 45, 53.
30 Newby, 1977, 161; Gasson, 1974a, 90.
31 Pedley, 1942, 65; Holderness, 1985, 104.
32 Ashby and Smith, 1938, 20.
33 Armstrong, 1988, 182–3.
34 Ashby, 1933, 229; Armstrong, 1988, 194.
35 Armstrong, 1988, 194, 195–6.
36 J. Wrigley, 1946, 76–9.
37 *The Hop Pocket*, January 1944.
38 Murray, 1955, 289–90.
39 Lund, Morris, Temple, and Watson, 1982, 10.
40 Armstrong, 1988, 229.
41 Havinden, 1966, 176.
42 Armstrong, 1988, 229–30.
43 Pedley, 1942, 13.
44 Mejer, 1951, 104.
45 Hughes, 1957, 35; Newby, 1977, 172.
46 Bessell, 1972, 37.
47 Compared to all industries and services the ratio of gross weekly earnings moved from 75 (1970–3) to 82 (1974–80), according to Lund, Morris, Temple, and Watson, 1982, Appendix X, 50.
48 Brown and Winyard, 1975, 5, 6, 14, 21.

49 Bosanquet, 1950, 80; Pedley, 1948, 108; Office of Population Censuses and Surveys, 1978, 77, 91.
50 Keatinge and Littlewood, 1948, 283.
51 Mollett, 1949, 223, 226; Health and Safety Executive, 1982, 5, 41.
52 In 1939 approximately one-third of all houses were new, nearly 4 million having been built since 1919. See Burnett, 1978, 242–3.
53 Pedley, 1942, 82, 88.
54 Agricultural Economics Research Institute, 1944, 123–4, 126.
55 Pedley, 1942, 82, 90; Shears, 1936, 4–5.
56 Pedley, 1942, 94.
57 Ministry of Agriculture, Fisheries and Food, 1967, 26; Gasson, 1975, 61.
58 Gasson, 1975, 81; Newby, 1977, 189.
59 See, e.g., Wise, 1931, 35, 41, 46, 153; Pedley, 1942, 112–18; Burton, 1943, 24–30, 30–46, 51–8.
60 Compare, e.g. Olive, 1938, and the ensuing discussion, especially 25, 30–1 with the official views of the Board of Education, and M. K. Ashby, as reported in Pedley, 1942, 119.
61 Pedley, 1942, 128.
62 Bosanquet, 1950, 83–90.
63 *Times Literary Supplement*, quoted on cover of the 1960 Penguin edition.
64 Newby, 1977, 218; Robertson Scott, 1926, 56–7; Armstrong, 1988, 185.
65 This is discussed by Armstrong, 1988, 187–90; Newby, 1977, 221–6 and, in more detail, by Howkins, 1985, 154–75.
66 Howkins, 1985, 160, gives the higher estimate. The figure for the county labour force is from the census returns for Norfolk, 1931.
67 Madden, 1956, 145–7; Newby, 1977, 235–6.
68 Pedley, 1942, 168. Note though, that perhaps as many as one-third had been in membership at one time or another (137).
69 Newby, 1977, 228.
70 See Armstrong, 1988, 220–1.
71 Madden, 1956, 277–9.
72 *Victoria County History, Staffordshire, VI*, 1979, 146; *The Hop Pocket*, April 1968.
73 Armstrong, 1988, 243.
74 Pedley, 1942, 1.
75 This is the opinion of the best-informed writers. See, e.g., Ashby and Evans, 1944, 84; Mejer, 1949, 38; Britton and Smith, 1947, 206–8; Nalson, 1968, 120; Newby, 1977, 200.
76 Newby, 1977, 236.
77 ibid., 371, 378, 387–8, 408.

# 9

# The decline of the country craftsmen and tradesmen

C. W. Chalklin

During the twentieth century the traditional craftsmen and tradesmen in the villages and country towns have become fewer and some types have almost disappeared. These craftsmen include those whose work is connected with the horse, that is, blacksmiths and saddlers, and makers of basic personal goods, such as tailors and shoemakers; men in old building trades, such as thatchers and handmade brick and tilemakers; manufacturers of household or garden or farm products, such as potters, turners, furniture makers, coopers and hurdlers, basket and hoopmakers; those involved in the processing of animal skins such as tanners, fellmongers, and, in some districts, glovers, and those connected with the food and drink trades, such as maltsters, brewers, millers, and butchers with slaughter houses. It is a nationwide phenomenon and the result of fundamental economic and social change.

The decline was already evident at the end of the nineteenth century, and the trend has continued since. Although it is indeed a general trend, the reasons for the decline of the individual industries can nevertheless be traced. Among the processors of agricultural products local tanners have disappeared. About 1850 there were seventy-five tanyards in Devon and Cornwall, but only six were active in the 1960s. The industry has become concentrated in Lancashire and Cheshire, the Midlands, the London area, and the old West Riding of Yorkshire, using not English oak bark but concentrated imported extracts, and employing machinery in large works.[1] Brewing and malting have become concentrated; and there are now many market towns where no malt is produced any more, as in the west of England. In the east malting developed, for example, on the outskirts of Sleaford in Lincolnshire between 1902 and 1906: there beside the railway, Bass built eight malt houses.[2] The small brewers in the country towns tended to disappear as large companies with economies of scale produced nearly all the beer: this tendency was already under way in the last quarter of the nineteenth century. Most noticeable is the decline of the windmill and water-mill. The number of windmills fell from about 350 in 1919 to 50 in 1946, and 21 in 1967. R. Wailes emphasizes the decline in dramatic fashion:

> In the Chatham and Rochester districts 80 or 90 years ago there were upwards of 50 working windmills; there are now none. In 1926 there were 48 working corn mills in East and West Suffolk alone; in 1939 there were 14; today (1971) there are but four, and only the tower mill at Pakenham, the mill familiar to millions on the television screen, is working with four sails.[3]

At the end of the nineteenth century the large flour mills at the ports began to take over much of the milling as most wheat became imported, and in the twentieth century motor transport has allowed easy dispatch to all areas. Trained millers prepared to work the mills have become difficult to find, while the disadvantages of relying on wind have also contributed to their decline.

Blacksmiths and saddlers have been hit by the decline of the use of horses, both on farms and in local transport. Horses used in farming declined from 802,044 in 1892 to 132,481 in 1954, while the total number of horses had fallen by a fifth in 1954 (1,169,146 in 1892; 218,534 in 1954).[4] Tractors have taken over on farms, a trend accelerated by the First and Second World Wars, and motorized transport began to become more widespread in the 1920s and 1930s. Saddlers became an aged occupational group, with very few young recruits, and many country saddlers' shops closed. While riding is a popular pastime in many areas, saddles are normally made in factories, though there is a continued demand for harnesses, collars, and repair work. Blacksmiths have continued to operate in smaller numbers by diversifying their work into agricultural machinery, such as tractors and harvesters, using up-to-date equipment and bigger workshops. In 1954 the Rural Industries Board reported:

> the rural workshop which for generations has relied on the forge and the anvil, now requires more equipment and a whole range of new skills if modern farm machinery is to be properly maintained and overhauled. More than twelve years ago the Bureau introduced gas welding instruction to rural craftsmen. This was followed by welding [using electricity], and more recently bench fitting, and the basic principles of good workshop practice have been introduced because it is impossible to repair modern agricultural machines with the necessary degree of precision without first raising the standard of these techniques.[5]

There has also been a continued demand for high-quality wrought ironwork which a few smiths specializing in the work have undertaken.

Among other crafts rural potters have declined, but the demand for unglazed horticultural pottery remained after World War II, made using electrically-powered equipment in modernized workshops.[6] With the decline of the making of wooden wheels (having been replaced by rubber-tyred metal wheels), wheelwrights have taken on a variety of jobbing carpentry, with a wide range of repair work. Since the Second World War, their workshops have become electrified. Some regional differences have existed and continue; in the north the wheelwright undertakes any work that is available; K. S. Woods noted in 1949 that:

> In the north where the wheelwright is also a joiner, doing any local work that comes to hand, it is still common to buy local trees for conversion. One Yorkshire wheelwright went back during the War to hardwoods and is specialising in machine-made saddle trees . . . he combines this with wheelwright's repair work.

Another Dales wheelwright at the same time made hay-rakes, while another made carts.[7] There are still woodworkers in the country making furniture by hand, turners, and makers of small boats and carts and trailers. In the underwood industry hurdles, agricultural and domestic fencing, and handles cleft by hand are still made, together with ladders and hop-poles. All this is of course on a small scale; wooden fencing has been replaced largely by barbed wire, and handles and ladders are normally machine-made. In the middle of the century small groups of craftsmen were working in the southern and midland counties, using coppice wood and thinnings, making forty or fifty products. In 1952 there were about 300 rural basket-making firms, but these were under competition from factory-made baskets using synthetic materials.[8]

Many small brickworks were closing down in the late nineteenth century and early twentieth century in the face of competition from machine-made bricks from the east midlands. Nevertheless there has been a steady demand throughout the twentieth century for good quality handmade bricks, used especially in public buildings such as schools. Labour-saving machinery has been introduced for extracting and processing the clay; improved drying equipment is used and kilns have been modernized, involving reduced coal consumption and more controlled heat.[9] Thatching ceased to be popular in the Victorian period on account of the fire risk and its suggestion of poverty, being replaced by machine-made tiles. The number of thatchers consequently fell.[10] But the picturesque appearance and the great advantages of rendering a house warm in winter and cool in summer have given thatchers work since the Second World War. In 1952 there were about 750 full-time roof thatching businesses, nearly all south of a line from the Wash to Swansea.[11] Building is still done by country-based firms, mostly employing a few men.

The decline of local tailoring and shoemaking is the result of the availability in the town shops of cheap, factory-made products. At the end of the nineteenth century many people still had their boots and shoes made by a local bootmaker, who carefully measured the foot and made a model for future use; but in the early twentieth century this ceased to be attractive with easy access to town shops and 'an enormous improvement in the style and shape and variety of fit in factory-made shoes'. Many country shoemakers consequently became cobblers in the early twentieth century, while local production in small workshops and factories supplied luxury articles or a special need.[12] Bespoke tailoring continued in a small way as a dying craft in Yorkshire (if not elsewhere), however:

> It was possible to see a tailor sitting cross-legged sewing on a board at Haws, Wensleydale, up to the 1960s. Bespoke tailoring, like bespoke shoemaking, still continues. We have talked chiefly to the Murphys of Richmond, and Mr A. Metcalfe and Mr E. Thompson of Haws who remember the craft, its jargon and lore.[13]

Studies of country parishes and country towns illustrate these national changes at the local level. The village of Gosforth in West Cumberland was the subject of a survey by Dr W. M. Williams in 1950–3.[14] This was a relatively isolated northern district where the dependence on locally made products was probably greater than in most parts of England. In 1900 Gosforth had three blacksmiths and many village craftsmen, including a basket-maker, clog-maker, watchmaker, and saddler, using mostly local materials. A tailor and dressmaker and sewing women mostly dealt with clothing, while bread was baked in the home. By the 1950s most of the craftsmen had gone. Clothes were all factory-made outside the district, and the villagers bought bread from the shops. There were still, however, a local mason, a joiner, and a wheelwright.[15]

A more recent history of Sedlescombe in Sussex is particularly interesting because of some of the personal details it includes, though it is not a statistical study. The tanyard was one of the first old businesses to close: it had gone by the 1890s. The other craftsmen gradually disappeared, some of them after the Second World War.

> In 1880 there were still three blacksmiths, a wheelwright, a carpenter's yard, a bootmaker and a bakery, besides two grocers-and-drapers, a toyshop, and a builder's yard. Even after the Second World War there remained the bakery, the blacksmith, the butcher and the builder's and the undertaker's, as well as the newsagent-post office and two grocers and greengrocers of today.[16]

The butcher continued until he died in 1956. The bakery and the builder's premises became antique shops; the last shoemaker finished in 1955; one brickyard closed during the First World War, the other in 1936. The first commercial garage opened early in the century in an old baker's and carpenter's workshop. In the course of time, the Bridge Garage had a fleet of Austin cars for weddings, funerals, and other taxi services, and the garage is still operating.

Commercial directories are an easily available source for tracing the decline of the country craftsmen, although the lists in nineteenth-century directories are probably less complete than those of the twentieth century. They were used by Professor Saville in a study of eighteen parishes in the South Hams region of Devon published in 1957.

**Table 9.1** The decline in rural craftsmen: 18 parishes in the South Hams

| | *1878* | *1910* | *1939* |
|---|---|---|---|
| Shoemakers | 43 | 23 | 9 |
| Carpenters and builders | 31 | 31 | 21 |
| Blacksmiths | 33 | 27 | 14 |
| Masons | 36 | 16 | 4 |
| Tailors | 21 | 9 | 2 |
| Wheelwrights | 13 | 9 | 8 |
| Thatchers | 17 | 3 | 3 |
| | 194 | 118 | 61 |

Source: Saville, 1957.

In this area the number of shoemakers and tailors reduced sharply under the competition of factory-made products. The similar fate of masons and thatchers presumably reflects the end of stone building and roofing with thatch, although some building did continue. Taking into account a decline of about 20 per cent in the population the figures suggest that wheelwrights and blacksmiths remained active until the Second World War.[17]

Finally, the present writer has analysed the Lincolnshire commercial directories for 1896 and 1933 to trace the extent of the decline in country craftsmen in twenty-five parishes. There was again a decline in the local making of boots and clothes, work on boots being restricted to repairs. Wheelwrights remained numerous, and the 1933 *Directory* named more saddlers and blacksmiths than did the 1896 *Directory*. The number of millers declined, there were no tanners at either date, and the one brewer of 1896 had gone by 1933. The building trades remained moderately busy.

At the end of the nineteenth century market towns were full of a variety of craftsmen. In Farnham, Surrey in 1890,[18] the makers of clothes and shoes and supplies of clothing materials were still strongly represented, with 12 dressmakers, 5 milliners, 7 drapers and a travelling draper, 6 tailors, and 7 shoemakers. There were also 5 watchmakers, saddlers were well represented (5), and there were 3 blacksmiths. In Horncastle (Lincolnshire) brewers and maltsters were quite numerous (8 and 6 respectively); there were 7 dressmakers, 4 milliners, 11 tailors, and 17 shoemakers. Trades linked with horses included 4 saddlers and 4 blacksmiths, and there were also 5 millers and 21 wheelwrights. These tra-

**Table 9.2** The decline in rural craftsmen: 25 parishes in Lincolnshire

| | *1896* | *1933* |
|---|---|---|
| *Agriculture and transport* | | |
| Blacksmiths | 11 | 18 |
| Saddlers | 1 | 5 |
| Wheelwrights | 11 | 12 |
| Agricultural implement makers | 0 | 2 |
| Threshing machine owners | 0 | 1 |
| Tractor owners | 0 | 1 |
| Cycle agents and repairers | 0 | 4 |
| Motor engineers | 0 | 3 |
| *Food processing and retailing* | | |
| Millers | 10 | 6 |
| Bakers | 7 | 5 |
| Butchers | 6 | 9 |
| Brewers | 1 | 0 |
| *Building* | | |
| Builders | 3 | 1 |
| Carpenters | 11 | 7 |
| Joiners | 0 | 3 |
| Thatchers | 1 | 0 |
| Bricklayers | 0 | 2 |
| Painters | 2 | 0 |
| Plumbers | 0 | 1 |
| *Miscellaneous* | | |
| Fellmongers | 0 | 1 |
| Tanners | 0 | 0 |

Source: *Kelly's Directory of Lincolnshire*, 1896, 1933.

ditional trades and crafts remained strong during the inter-war years. The 1933 *Directory* records no less than 17 blacksmiths, 12 wheelwrights, and 4 saddlers. On the other hand, those drawing a livelihood from the making of clothes and shoes had nearly disappeared; the shoemakers had now become shoe repairers (6); there were no dressmakers, and only 2 drapers. There were newer trades, however, such as cycle dealers and repairers (4), a tractor owner, and motor engineers. A relatively recent *Directory* of Farnham, Surrey (1971), now a much expanded commuter town and shopping centre, is illuminating in respect of the changed character of the occupations. While there were few drapers, only 2 tailors and a single dressmaker, there were no less than 40 grocers and 10 greengrocers, 14 butchers and 5 fish and chip shops. The town was full of trades characteristic of affluent contemporary towns: 2 antique dealers, an antique restorer, 3 booksellers and an antique bookseller, 19 ladies' hairdressers, a variety of trades connected with motor vehicles, including 11 motor engineers and garages and 9 motor car dealers, and also 9 television dealers and 18 electrical firms.

Alongside the decline of the traditional crafts new industries have been set up in the country towns and to some extent in the rural areas. This has occurred particularly since the Second World War. Labour (especially that of women and girls) is usually flexible and cheap, land is inexpensive and rates are low, and because of motorized transport, firms can enjoy a large catchment area for workers. Since the report of the Scott Committee in 1942 employers have often decided that the advantages of rural industrialization outweigh the disadvantages. It has been planning policy to site factories in country towns rather than villages or the countryside, but there has been much successful encouragement of industry in rural areas, too. Good communications are usually important in the choice of a region by manufacturing firms. A few of these industries are large employers of labour, such as the Canadian Northern Aluminium Co. started at Banbury in 1931: by 1954 there were 2,300–500 workers in its laboratories, and rolling, sheet, and extrusion mills producing alloys of aluminium for many industries. Between 1927 and 1954 eleven light industries were also established at Banbury. Westland Aircraft came to dominate employment in Yeovil. Didcot's population grew five times between 1931 and 1954, largely on acount of the ordnance depot.[19] In some areas the factory processing of farm produce has become important, such as sugar beet in East Anglia and the processing of cheese and other dairy products in parts of the West Country. In other regions such as Cornwall and north Somerset quarrying is important.

The typical country town has a variety of light industries. This may be illustrated by the case of the small town of Steyning in Sussex (population 1,885 in 1931, 3,284 in 1971): in 1958 there was a firm of power-farming contractors, and in 1976 agricultural engineers; other businesses in 1976 included a firm of motor engineers, and since the Second World War the chief employers in the town have been the firm of F. Duke Ltd, builders, decorators, and timber merchants, founded in 1902, and employing about 90 people in 1958, and a firm manufacturing precision gramophone equipment, which in 1976 employed about 140 people. In the 1930s there was a Steyning Lime and Brick Co., and chalk was still being extracted in 1975. Pianolas were made after the Second World War. Typical public employers included a waterworks company founded in 1897, a fire station, an electricity generating station begun in 1914 but closed by 1948, and the gas company, which ceased working in 1958.[20]

A few country towns have greatly expanded their market function; others have lost it as motorized transport by lorry and car have made closely sited markets superfluous. As A. Raistrick wrote in *The Pennine Dales* in 1968, 'a farmer's market radius has increased. No farm is too remote now from a market and sheep and cattle are quickly transported over long distances in a comparatively short time and in good condition. Farmers in Airedale can quite easily attend markets in Hexham or Tynedale if they wish, and some of them do so.'[21] Markets tending to concentrate trade are those which offer extensive covered accommodation, up-to-date facilities, and excellent road links. Large market auction companies carry on a great trade in livestock. At Cockermouth in Cumberland a company formed in 1873 and run by Robinson Mitchell had 11,537 cattle, 61,042 sheep, 284 pigs, and 1,316 horses passing through the market in 1876; in 1978 the numbers were 12,000 store cattle, 24,000 store sheep, 450 dairy cattle, 7,500 calves, 3,250 fat cattle, 63,500 fat sheep, and 1,350 fat pigs. The 1981 calendar features a market every Monday for fat stock, dairy cattle, and calves, the fortnightly sale of store cattle, cows, and store sheep, and annual sheep sales in the autumn.[22] In Devon and Cornwall 'markets have closed not only in remote villages but also in such old towns as Crediton, Okehampton and Bodmin. On the other hand, road transport has reinforced the vitality of certain town-markets and road junction sites in the large area west of a line from Bideford through Okehampton and Tavistock.'[23] This emphasizes the fact that the small historic market town has remained active: in Oxfordshire the markets of Chipping Norton (population 3,879), Thame (3,585), Wallingford (3,514), Farringdon (3,500), Woodstock (only 1,713), and Bampton (only 1,279) were still in use in 1954. But 'there are also several small towns whose markets have declined or disappeared in modern times or whose functions as centre for the surrounding villages have largely ceased':[24] towns with populations of up to 1,500 such as Deddington, Burford, Dorchester on main roads; and less accessible, Charlbury, Eynsham, Islip, and Watlington.

From the point of view of employment one should not overemphasize the importance of markets. Most employment in these towns is in public and private consumer services, and in retailing. They provide schools, medical, and social services, and a variety of increasingly attractive shops with an ever-widening range of goods.

Tourists and commuters have transformed some country towns and affected the character of many more. Antique shops and teashops are important amenities for tourists in the historic towns. In the countryside farms provide camping sites, 'farm teas', and roadside shops, as well as farmhouse holidays. Commuters live in country towns and villages because life is pleasant there. London commuters have greatly enlarged the market towns of the home counties and filled many homes and cottages in the countryside; and other English cities have had a similar, though of course smaller effect on the neighbouring country towns and villages. Trains and motor cars have made travel to work a normal practice. Thus of the village of East Meon, Hampshire, twelve miles north of Portsmouth, with a population of 1,000 in 1981, F. G. Standfield wrote in 1984:

> Of persons employed within the parish, farm workers still constitute the largest occupational group, and those self-employed include shop-keepers, a builder, electrician, heating engineer, and baker (the latter also owning the part-time fish and chip shop). Most wage-earners travel daily, usually moderate distances, to employment outside the village in

> the service of local authorities, public utility and transport undertakings, hospitals, education establishments, shops, offices, surgeries and building contractors. Likewise members of the professions, of the armed forces and of the merchant navy obtain their livelihood beyond the parish.[25]

Many villages have attracted retired people and those wanting second homes as the number of people of pensionable age has grown and wealth has spread among a greater proportion of the population. This does not necessarily mean that villages have ceased to be communities, as their social activities are often considerable, but agriculture and its connected trades no longer dominate village life. Particularly since the Second World War the market towns and countryside in many parts of England, and not only in the south and the midlands, have become more prosperous as government and cheap labour have encouraged the siting of modern industry, especially light manufacturing, in these areas; as the Rural Industries Board and consumer demand have kept many of the traditional crafts alive, helped by improved technology; and as tourism and especially commuting have become of major economic importance.

## Notes

1 Shorter, Ravenhill, and Gregory, 1969, 158.
2 Brown, 1986, 78.
3 Wailes, 1954, 179–80.
4 W. M. Williams, 1958, 207.
5 Rural Industries Bureau Report, *Country Craftsmen and Rural Industries*, 1954, Trustees of the Rural Industries Bureau, London, 7.
6 Rural Industries Board Report, 1949–50, 17.
7 Woods, 1949, 54.
8 Rural Industries Bureau Report, *Skill in Country Workshops*, 1953, Trustees of the Rural Industries Bureau, London, 25, 27.
9 Rural Industries Bureau Report, 1950, 14–16.
10 Wymer, 1946, 49.
11 Rural Industries Bureau Report, 1953, 18, 21, 22.
12 Wymer, 1946, 89–90.
13 Hartley and Ingilby, 1968, 103.
14 Williams, 1956.
15 ibid., 20–1.
16 Lucey, 1978, 346, 360, 368, 376, 384, 386.
17 Saville, 1957, 176, 212.
18 See *Kelly's Directory of Farnham*, 1890.
19 Martin and Steel, 1954, 145, 155.
20 *Victoria County History: Sussex VI*, 1980, 237, 239–40.
21 Raistrick, 1968, 135.
22 Bradbury, 1981, 108.
23 Shorter, Ravenhill, and Gregory, 1969, 187.
24 Martin and Steel, 1954, 145.
25 Standfield, 1984, 133–4.

# 10

# The new culture of the countryside[1]

Michael Winstanley

Mourning the passing of old, 'traditional', country ways of life remains highly fashionable. Modern farming is condemned for despoiling the landscape and denying habitats for wild life and pleasure for residents. Small or scattered settlements which once blended with their environment are portrayed as festering sores swollen by 'newcomers' with their urban attitudes and habits. In these circumstances the concept of a specifically 'rural' way of life seems increasingly irrelevant, swamped by an all-pervasive metropolitan culture which is obliterating regional distinctiveness and the rich variety of local customs and practices, submerging them under a tide of uniformity, their purity and authenticity 'contaminated' and destroyed by corrupting consumerism.

Such views are not only expressed throughout the popular media today but, less sensationally, they underpin sociologists' analyses of the contemporary countryside. Prior to the 1960s works had been replete with references to an urban-rural divide or continuum, to the contrast between the seemingly anonymous existence of the city (*Gesellschaft*) and the apparently contented intimacy of rural small-scale communities (*Gemeinschaft*).[2] Such ideas, which owed much to the American school of rural sociologists, are no longer in vogue. Ray Pahl's much quoted phrase encapsulates the change of outlook: 'In a sociological context the terms rural and urban are more remarkable for their ability to confuse than for their power to illuminate.'[3] Only in terms of land use is 'rural' now considered to be a meaningful term; the study of rural 'communities' has been replaced by the study of 'local social systems' which do not assume that 'rural areas' are endowed with a qualitatively different way of life.[4] Settlements are '*urbs in rure*', 'discontinuous suburbs', 'metropolitan villages', distinguishable from their urban counterparts only by being surrounded by an open countryside which has no significant impact on their physical, cultural, or spiritual well-being. Chroniclers of the process have coined a variety of catchy phrases to describe it: *The End of Tradition, The Quiet Revolution*, or, quite simply and with echoes of a previous age, 'Change in the village'.[5]

These developments are usually portrayed as being of comparatively recent origin, the product of a post-war social revolution which has transformed the cosy world which Agatha Christie's Miss Marple inhabited into one where P. D. James's cosmopolitan Inspector Dalgliesh is more at home. The pace of change since the 1940s is certainly impressive but concentration on obvious, dramatic discontinuities at the expense of a longer term perspective can obscure both earlier influences on rural society and exaggerate the extent to which it once functioned as a 'community'.

The erosion of the 'real' country style of life has long been a fashionable subject in literary and historical circles. V. Bonham-Carter commented sadly that in post-war Britain only 'the relics of a regional culture' remained: isolated sword dances, mumming, furry dances, beating the bounds.[6] Over a hundred years earlier, however, in 1838, William Howitt was already commenting on much the same thing:

> How rapidly is the fashion of the ancient rural life of England disappearing! . . . How many of the beautiful old customs, the hearty old customs, the poetical old customs, are gone! Modern ambition, modern wealth, modern notions of social proprieties, modern education, are all hewing at the root of the poetical and picturesque, the simple and the cordial in rural life.[7]

Over the following fifty years the surviving elements were subjected to waves of assaults from transport improvements, population decline, the collapse of landed patronage, agricultural change and depression, advancing literacy, and the corrosive influence of compulsory education. By the end of the century Cecil Sharp and his folklorists were really collecting evidence of England not as it was but as it had been – or how they wanted it to have been. In this cultural malaise, as in more material aspects, the countryside was seen to be in crisis, 'the rural problem' dominating academic, popular, and polemical literature of the time. Recent studies have confirmed many of the fears and concerns of these commentators; the countryside which this century inherited was one which had already lost much of its cultural identity. 'The triumph of an increasingly ascendant urban culture was occurring almost everywhere', concluded Robert Malcolmson recently.

> And the result was that leisure in the countryside became little more than a modified offshoot (often in a paler form) of the recreational culture of the cities. A distinctive rural recreational culture, readily identifiable and central to the life of the country labourers, had been largely extinguished.[8]

It is important, too, to appreciate that the distinction between 'urban' and 'rural' ways of life has also been blurred in the past by migration to the towns. Just as 'newcomers' to the countryside now bring with them their own practices and habits, so migrants to the towns in the last century brought with them their own mutual support networks, superstitions, songs, medicinal remedies, and a dialect as a way of coping with the strangeness of the new environment. What we might loosely call pre-industrial values and life-styles have often thrived precisely where we would least expect to find them. Studies of cotton Lancashire, for example, the cradle of the industrial revolution, have increasingly stressed the dogged persistence of practices previously associated with the mixed domestic farming economy of the region. In these towns the textile manufacturers had to contend with disruptions to production occasioned by the celebration of the wakes, annual holidays whose original purpose, the relaying of rushes on the church floor on the local parish saint's day, was more honoured in the breach than in the observance. Nevertheless, elaborate rush-carts continued to be drawn through the towns by gangs of young men until well into the closing decades of the last century, much to the annoyance of some of the more 'respectable' inhabitants who belittled such 'rustic merriments'.[9] This century has seen yet more manifestations of ruralism in an urban context, most visibly in the domestic architecture of the inter-war years with its rustic charm, and in the popularity of gardening.[10] It is not only in the economic market-place that town and countryside have enjoyed a symbiotic interdependence; the trading of goods has long been paralleled by the trading of cultural influences. Although there may well be evidence of a 'common material civilisation and of a common culture', as Arthur Ashby astutely observed in 1939, 'neither the civilisation nor the culture was entirely urban in origin and not all changes begin in towns'.[11]

While all this qualifies some of the emphasis placed on the recent origins of the decline of a distinctive rural way of life, it does not undermine the basic premise of such studies which argue that a significant convergence of life-styles has become more evident in the last fifty years. The most striking feature of most rural areas today is just how little the presence of fields, woods, hedgerows, livestock, and crops is responsible for moulding the daily lives of the majority of those who live in their midst. The life-style that is found in the country today is not *of* the country; fields may provide the scenic backdrop but they no longer determine the action on the stage. The seasonal cycle of agricultural life which once dominated the social calendar is increasingly irrelevant since, as earlier chapters have shown, only a small and declining percentage of country dwellers have direct connections with the land, and these, if they are not corralled into council houses, increasingly reside away from the villages on farms or in isolated farm cottages. The broader culture, as in the urban context, is virtually divorced, often geographically, from the workplace, compartmentalized into a separate sphere.

This divorce from the land can be illustrated in a wide variety of ways. Rural publicans, once intimately involved in the local economy as small farmers, artisans, carriers, or simply suppliers of accommodation for facilitating the exchange of information and produce, have either closed their doors or cater for a very different customer, the car-borne visitor in search of authentic country surroundings, 'ploughmen's' lunches, or the real ale advertized in CAMRA publications.[12] Local dialect has fallen into disuse as the tools, jobs, and routines to which its vocabulary referred have disappeared and as mass communication and education have imposed standardized language usage. It is more likely to be found in the pages of local dialect dictionaries or the scrapbooks of Women's Institutes than on the lips of local people. Herbal medicines and folk remedies have been similarly displaced, in this case by a reliance on professional experts in the shape of doctors and veterinary surgeons with more 'scientific' training and methods, or in the case of the waggoner with his private horse, magic by the mechanic.

Even the country school's curriculum has lost its 'rural' bias. Up to the 1930s teachers were encouraged to stress gardening, horticulture, livestock management, and other practical skills as part of a suitable training for agricultural work. As openings here disappeared the Board of Education's pamphlet of 1934, 'Education in the Countryside', shifted the emphasis simply to inculcating in pupils a greater appreciation of their surroundings. Since the war, however, even this has been superseded by a concentration on more general skills to assist geographical and social mobility.[13] With the gradual implementation of the 1944 Education Act and the abandonment of all-age schools in favour of separate secondary education, usually in an urban context, the scope for education to reflect and preserve any values and skills specific to the countryside has been further restricted. Where it has survived, 'rural studies' has tended to be viewed as more suitable for lower ability children, although even this is likely to be threatened by the national curriculum which is now to be used in all state schools.[14]

There is little evidence either of any rural bias in most of the formal activities and societies which exist in the country, apart from the obviously pragmatic ones like the National Farmer's Union, whose members share a common economic interest. The failure of its counterpart, the National Union of Agricultural Workers, to retain a separate existence, and its reluctant merger with the Transport and General Workers' Union, are symptomatic of the declining feasibility of organizations appealing specifically on agricultural grounds. Even Women's Institutes, founded after World War I to enhance the quality of agricultural life, have gradually shed their original purpose, lecture programmes and activities now reflecting matters of general rather than purely local and rural interest.[15] The overwhelming majority of societies which flourish in rural areas are manifestations of a broader cultural pattern and often function as branches of national movements whose origins, inspiration, and finances have their roots elsewhere. The same is true of local sports and pastimes, except those which, like fishing and shooting, have to be pursued in the countryside. Even where examples of local sports, such as bat and trap or stool ball, have survived they have been overshadowed by the introduction of other pursuits like football with nationally agreed rules and procedures. National holidays have also superseded local celebrations and seasonal markers like Michaelmas. Even children's activities seem to be largely divorced from the natural environment with their increasing dependence on home-based, commercialized pursuits and games.[16]

This divorce of culture from the land is hardly surprising since most of those who now live in the country do not work there. It has, of course, long been the practice of successful businessmen to retire to the country but larger-scale migration has been possible only as general living standards have risen and transport facilities have improved. Isolated rural suburbs appeared around railway stations from the late nineteenth century, a process which picked up momentum in the inter-war years. In the post-war decades, however, the internal combustion engine has opened up virtually the entire country to colonization by the urban middle class.[17] Over most of England, especially the Home Counties, the invasion has taken the form of commuter or retirement homes. In the more geographically isolated areas like mid Wales and the Lake District where permanent settlement is not possible due to distance from centres of employment, second homes have increased dramatically, creating eerie deserted villages for much of the year.[18] The arrival of these newcomers has prompted further concern about the quality and vitality of rural life.

The decline of 'community', the seemingly inevitable consequence of this migrant invasion, is a recurrent theme in literature on the countryside. The phrases used to describe this lost world are remarkably consistent. It was 'pure', 'natural', even 'organic' (a word increasingly in vogue), and contrasted favourably with the privatized, polarized, unnatural life-style of today. It was distinguished by 'a vigorous social life at both formal and informal levels', by being 'socially integrated', with an 'innate sense of belonging', where 'knowing one's place' implied acceptance of, if not contentment with, what is invariably portrayed as a hierarchical social order.

There are enormous problems in assessing the accuracy of these statements. 'A sense of community', after all, is entirely subjective, and is so dependent on the age, sex, social class, and value judgements of the researcher or participants that sociologists have now abandoned the concept as 'no longer operationally defineable'.[19] Measurement of such feeling, difficult enough in the present, is virtually impossible for the past. The 'traditional' rural society that possessed this quality is also all too frequently assumed in the literature to fit to the hierarchical, 'closed' village model, dominated by squire and church. It is significant, perhaps, that the self-governing, more egalitarian rural society which dominated much of the uplands of the north, or the freer open villages which lacked big houses and clerical edifices, features less often in the literature.[20]

Most agricultural 'communities' were hardly egalitarian or utopian. Landowners and clergy, who tended to view themselves as linchpins and who are often prominent in expressing regret at the passing of such places, were in many respects outsiders, enjoying wider involvement in what they freely called 'society' but largely excluded from the daily, informal activities of the poor who made up the majority of the population. Any mutual assistance which existed was born out of necessity, and the security which knowing each other's business' brought went hand in hand with an absence of privacy and a stultifying conformity of behaviour embodied in unwritten rules of conduct. These were also communities riven with class distinctions. Upward social mobility was rarely achieved except by migration. Rigid social distinctions were taught to children. 'There was a tremendous difference between different classes', recalled one vicar's daughter who was not allowed to play with local children. 'You were allowed to be one class or another . . . everybody was divided up into very very strict categories.'[21] Even seemingly small market towns had their distinct residential segregation and classes rarely mixed.[22] In reality then, many rural communities were characterized by intolerance, social rigidity, and restrictiveness, and poorer members were prisoners, locked in by walls of ignorance, immobility, and lack of opportunity. To imply, as some do, that such a situation was 'organic', 'natural', or 'traditional' is a travesty. It was a product of wider commercial forces and exploitation.

The stereotype rural community contrasts markedly with its counterpart in the towns. Although it has been argued that there were factory communities in Lancashire which mirrored the hierarchical structure of the closed parishes of the countryside, with cotton industrialists acting as 'urban squires', engendering feelings of dependence and deference,[23] the impetus for, and operation of the working-class community came primarily from below, as mutually supportive local social systems, but which, unlike the physically distinct rural village, remained until recently invisible and unrecognized by the outsider.[24] 'Every industrial city, of course, folds within itself a cluster of loosely defined overlapping "villages" ', recalled Robert Roberts of his Salford childhood. 'Those in the Great Britain of seventy years ago were almost self-contained communities.'[25] These too, like their rural counterparts, have acquired a nostalgic glow over the years as redevelopment has threatened their existence.

Rural society today fits neither the hierarchical nor the egalitarian stereotyped community. It is far from socially homogeneous and its class polarization corresponds more closely with length of residence than with a specific relationship with the land. Despite the fact that some long-established residents consider that social distinctions have actually declined and have retained their own mutually supportive networks, they still maintain that they experience a sense of alienation and loss because they no longer comprise the majority of the population.[26] They feel isolated, outnumbered, and often disadvantaged in villages which no longer seem to belong to them but to newcomers, some of whom seem unwilling to lay claim to possession. The latter group often appear distant, uninterested, preferring to see their flight to the country as a private decision, a search for peace and tranquillity, a retreat rather than a longed-for communal experience, and they resist identifying themselves too closely with their new places of residence. Such people exercise their ability to choose and look outside for work, friends, and leisure, expressing little desire to restrict their activities primarily to the immediate vicinity. They often feel that they do not need the support of others in their immediate locality to survive. These people are not regarded and do not regard themselves, as 'villagers'. In the central Surrey settlements examined by John Connell in the 1970s many interviewees positively recoiled at the suggestion that they might regard themselves as such: 'Villager conjures up a toothless old wreck by the village pub. No thank you . . . certainly not, sounds country bumpkinish, someone very parochial, not going anywhere.'[27] Involvement in local activities and identification with a restricted geographical locality for many of these rural dwellers is now voluntary and partial, not compulsory and total.

The proliferation of formal societies, however, might imply otherwise. Contrary to the expectations of their originators in some cases, social surveys have discovered that villages, even largely commuter ones, were far from being cultural or recreational deserts. The Surrey villages of East and West Horsley contained no fewer than thirty-five societies in the mid 1970s catering for all age-groups, music, drama, sport, gardening, politics, social welfare, hobbies or just general social gatherings and good neighbourliness. 'Local life', concluded Connell, 'thrives now as it has never done before.'[28] In his survey of WI scrapbooks in the 1960s, Paul Jennings highlighted Horncastle as a particularly lively country town with, *inter alia*, clubs for photography, art, coin collecting, archaeology, bowls, tennis, football, and rifle shooting (a marked contrast to the pubs and brothels which proliferated in the 'traditional' nineteenth-century town).[29] Surveys elsewhere show that such high levels of formal association are not unusual and that smaller rural settlements still tend to have a much larger number of such societies, relative to their population, than urban ones.

Yet existence of such societies does not mean that their members have a wider sense of belonging. Most of them have a limited, sectional appeal and rarely attract the poorer elements of the population. They represent at best a series of interlocking social systems, at worst evidence of the way in which rural society is fragmented along the lines of class, sex, and age. Only the Women's Institute stands out as a body aiming to attract all classes irrespective of age, but despite the fact that the number of branches has remained steady, membership has fallen significantly since the early 1970s.[30] Others enjoy only a short life, or are sustained by a small band of enthusiasts or by support from outside sources. Even the village hall or 'community centre' is all too often a middle-class concept and construction which fails to appreciate that it could be superfluous to local informal networks which require no architectural embellishments. Whether such networks really worked in the past, or exist today, and, if so, whether they encompass more than specific subgroups within society, is less clear. They appear to be strongest amongst the new arrivals on recently constructed estates or among the council tenants in essentially commuter villages, forming local social systems within the village. Identification with an area is also more evident amongst women who do not leave the village for work or who are involved in bringing up families, suggesting perhaps that more women are likely to equate the geographical with the social community.[31]

Although it still enjoys the largest single membership, the Church is now only one of the many formal organizations which exist in the country. Its changing role, and the upswing of its symbolic representation of community identity by the school, highlights another major change, the way in which secular concerns and organizations now dominate rural culture.

The clergy were once the most visible leaders of rural society, a position derived as much from their professional status and wealth as their religious role. Their authority has never gone entirely unchallenged and was far less complete outside the Church's 'heartlands' in the south and east. Attendance levels, however, have been persistently higher in the country than the towns, and probably rose from the late nineteenth century, despite the Church's economic problems, as strenuous efforts were made to modify services, become involved in the provision of education, local government, and voluntary associations, and to incorporate local customs into services.[32] In the countryside, observed Leslie Paul in his influential survey of the Church in 1964, 'The Church always seemed happiest and at its most socially useful' and it 'was socially as well as religiously meaningful to be a member of the Church'.[33]

The problems which the Church has faced were not solely caused by the rising tide of secularism; they relate to its inability to adapt to the shifting population distribution in the countryside. The parish structure has restricted its ability to change, livings have not always been in the church's gift, local endowments have tied income from investments to specific areas, the fixed capital of church and vicarages has needed maintaining and could not easily be disposed of. This has led inevitably to a mis-match of resources and needs. While those in expanding settlements struggled to cope, ten out of every eleven parishes in England by the early 1970s had populations of less than 3,000, and in the smallest of them the priest was often socially isolated and frustrated by the lack of work.[34] These areas, rather than the growing commuter settlements, are still viewed by the Church as its primary rural problem and there has been considerable debate about how to manage them. Before the mid 1970s it was common practice to allocate them to older men until the introduction of a compulsory retirement age of 70 curtailed this option. Consequently, church authorities have had to withdraw manpower from depopulated areas as the number of ordained priests continued to fall from nearly 15,500 in 1961 to just under 10,750 by 1984. Following the recommendation of the Sheffield report in 1974 to reduce the numbers of rural clergy by 50 per cent, the Church has resorted to a combination of reduced services, team ministries, the merging of parishes, and reliance on lay preachers, non-stipendiary ministers (both self-selected and nominated by congregations), lay elders, and pastoral assistants. Although some commentators are enthusiastic about the increasing involvement of members in the running of the Church and welcome it as evidence of 'community ministry',[35] these trends unquestionably represent a weakening of the Church's presence in such areas and raises the distinct possibility that it will suffer the same fate as its once great rival, Nonconformity, which relied on remarkably similar organization and lay involvement.

The problem in the expanding rural parishes is rather different. Here the clergy have to compete with the attractions of other groups and institutions; they no longer come close to being 'pivots of social life', the 'cogs around which village life revolved'.[36] There are now others, often professional immigrants like the vicar himself, who are eminently qualified to provide rural leadership. For those parishioners who are regular attenders, still drawn primarily from the higher classes, church duties and societies can continue to provide an all-embracing social life, but this religious community represents but a small fraction of the wider population and conforms to what Anthony Russell has called 'the gathered church model', abandoning the claim to cater for the wider secular community.[37] Other residents choose more selectively from a varied recreational menu, only some of which, especially mothers' unions, scouts, guides, and youth clubs, are attached to the Church. None of these has any overt religious content and the appeal remains largely social, so much so that some clergy seriously question their continued involvement in them. Their impact on Church attendance would appear to be minimal and every index of religious influence – baptisms, confirmation, numbers on the electoral rolls – shows little sign of recovery.

The new symbols of community identity are overwhelmingly secular and it is the school, decreasingly under religious auspices,[38] which has undoubtedly ousted the Church as the major institutional underpinning of village life. On one level this is unexpected. Schooling, even in the days when its curriculum was more closely related to the agricultural environment, was often regarded as having a disintegrating effect, encouraging young people to seek an elusive better life elsewhere. Farmers were not backward in their condemnation of such book learning, and absences in schools' log books indicate that parents' priorities did not always reflect those of the educational providers. Serious doubts have always been expressed about the quality of schooling provided in such schools: resources are stretched too thinly, children have restricted curricula, mixed age-groups hold back brighter pupils, and the low level of pay, especially before Burnham, attracted poor quality teachers. The rise of the village school's reputation also coincides with the decline in its educational importance, since children now spend less than half their school life there. And yet the school, where it has survived, is undoubtedly the most potent unifying force in rural areas today. Closure is invariably fought, the village unifying to fight outside interference, and receiving advice and aid from advisory bodies which issue leaflets on 'Community support for village schools' and 'Fighting a village school closure'.[39]

Why should this be the case? There has certainly been little in the way of official encouragement for anything other than a purely educational role, except from Henry Morris, who as Head of Cambridgeshire's Education Service for nearly thirty years from 1924, had a vision of a 'series of cultural communities' based around all-embracing village colleges which doubled as social and recreational centres. Largely as a result of intense personal commitment and outside funding he succeeded in several areas, but his example has not been followed elsewhere and some of his institutions no longer match his ideal.[40] Schools, from the perspective of the bureaucrat, are simply educational establishments; they can be closed by pointing to educational deficiencies, falling rolls, or excessive costs. The tenor of campaigns to prevent closure, however, rarely reflect these issues, except in considering increased transport costs. They lay overwhelming stress on the social roles which the schools fulfil, and employ language reminiscent of descriptions of the old 'organic' society. Closure is invariably portrayed as heralding the 'death of the rural community' since the school is 'one of the most vital organs', the heart which pumped life blood around it. The teachers are portrayed as linchpins in organizing local events. The school building is portrayed as the only place where public meetings can be held. A local education arguably imparts security for younger children, giving them a 'sense of belonging'. Through involvement in the school adults get to know one another, since it provides a focus both for formal and informal association.[41] Without a school families would cease to be attracted to an area, which would fill up with retirement homes and holiday cottages. Surveys of rural attitudes all point to the same conclusion: 'The sense of community identity felt by village residents . . . was found to be strongly correlated with the presence or absence of a primary school.'[42]

The full implications of the social revolution which has swept over the countryside in the last fifty years are not yet fully apparent, but it is clear that among them is to be numbered the final eclipse of anything remotely resembling a distinctively 'rural' way of life. 'Traditions' and customs are now 'rather self-consciously fostered', as Bonham-Carter scathingly commented, 'by certain educationalists and writers' or deliberately revived, often for commercial reasons, as quaint reminders of our 'heritage'.[43] It has been a social process which owed little, if anything, to deliberate attempts to control, direct, or repress, and much, as Arthur Ashby recognized, to the inevitable working out of market forces and improved channels of communication.

> It would be impossible to have two populations with radically different standards, different sets of attitudes and values, while to-and-fro travel is as easy as it is at present, while broadcasting and newspapers present standard sets of news and views, and while industrial transfer and residential migration are unrestricted except by the personal equipments of individuals and variations in their economic resources.[44]

We need not necessarily regret these changes. Even a hearty dislike of the 'new culture' should never allow us to forget the shortcomings of what went before. Those who remember the old village life, despite their misgivings about the present, are amongst the most enthusiastic supporters of those very things which allegedly have undermined it: better communications, shops, industrial jobs, and popular entertainment.

Yet the fascination with a lost rural arcadia shows no sign of diminishing. This is because its appeal, and the concern with preserving its 'environment' and 'heritage', span the political spectrum, although conceptions about what constitute 'natural' or 'traditional' aspects of country life differ greatly. On the one hand, it underpins the literature of radical groups which attack the capitalist power structures, over-zealous commercial farming, and the abuse of privilege, and champion the retention and restoration of common rights and customs.[45] On the other, and even more successfully, it has been the basis of the political right's deliberate invention of an idyllic semi-feudal past of benevolent squires, church spires, leafy lanes, and country 'sports', and the weaving of these images into a pictorial and literary tapestry which purports to encapsulate national identity and virtues.[46]

The characteristics of the vanished countryman – and woman – are as elusive and debatable as the concepts of a traditional countryside or way of life. Two hundred years ago, the Tory Somerset parson, William Holland, was in no doubt that the country-dwellers in his area were 'very slow and lazy and discontented and humoursome and very much given to eating and drinking'.[47] During Victoria's reign 'Hodge' was a dominant image, a besmocked, backward, ignorant buffoon who epitomized the idiocy of village life. Only after the 'rediscovery of rural England' around the turn of the century did he become endowed with qualities which previous generations never remarked upon: patience, forbearance, sensitivity, pragmatism, rugged individualism, and loyalty. The concepts of what constitute 'traditional ways of life' and true countryman are thus constantly being reformulated to meet the psychological needs of contemporary society. Whether the 'new' culture of the countryside will ever be looked back on with affectionate yearning as in some way 'traditional', remains to be seen, but the increasingly evident elevation of inter-war society to that rank, and the fierce loyalty already being manifested towards institutions like schools, tempts one to suspect that it will.

## Notes

1 I am grateful to my colleague, John Walton, for constructive comments on an earlier draft of this chapter.
2 Tonnies, 1887.
3 Pahl, 1966, 229.
4 Bradley and Lowe, 1984, 1–9; Newby, 1978, 3–30.
5 Ambrose, 1974; Connell, 1974; Connell, 1978; Pahl, 1964; Newby, 1979, chapter 5.
6 Bonham-Carter, 1952, 206.
7 Howitt, 1838, 582.
8 Malcolmson, 1981, 615; see also Howkins, 1973 for case-study.
9 Poole, 1985.
10 Burnett, 1978, 261–2.
11 Ashby, 1939, 353.
12 Hutt, 1973.
13 Rolls, 1965, 177–85.
14 Nash, 1980, 29–34.
15 Connell, 1978, 138–9; Bracey, 1959, chapter XIII; Jennings, 1965.
16 Ambrose, 1974, chapter 10.
17 Rogers, 1989.
18 Capstick, 1987; Bielckus, Rogers, and Wibberley, 1972.
19 Stacey, 1969.
20 Exceptions include, W.H. Williams, 1956; Littlejohn, 1963.
21 Interview with Mrs J. Wray, b. 1899.
22 See, for example, Brigg Local History Group, 1983.
23 Joyce, 1980.

24 Among the pioneering studies were Anderson, 1971a (on the nineteenth century); Young and Willmott, 1957.
25 Roberts, 1971, 16.
26 Newby, 1979, chapter 5.
27 Connell, 1978, 152–3.
28 Connell, 1978, 134.
29 Jennings, 1965, 196; Davey, 1983, 9–26.
30 *Social Trends*, 1987, 173. Membership was 400,000 in 1971, 350,000 in 1984; the number of branches rose from 9,155 to 9,190.
31 Stebbing, 1984.
32 Obelkevich, 1976, 103–82; Morgan, 1982, 167–74.
33 Paul, 1964, 46.
34 Beeson, 1973, 82–8.
35 Russell, 1986, chapter 12.
36 Crichton, 1964, 18; Ambrose, 1974, 88.
37 Russell, 1986, 267.
38 Bamford, 1965, 59.
39 Pamphlets available from Action with Communities in Rural England, 26 Bedford Square, London.
40 Ree, 1973.
41 Forsyth, 1984, 209–24; quotes from letters in *Lancashire Guardian*, 21 October 1987.
42 Shaw, 1978.
43 Bonham-Carter, 1952, 206.
44 Ashby, 1939, 354.
45 Shoard, 1980.
46 Howkins, 1986.
47 Holland, 1984, 16 (29 October 1799).

# References


Abrams, M. A., 1932, 'A contribution to the study of occupational and residential mobility in the Cotswolds, 1921–1931', *Journal of the Proceedings of the Agricultural Economics Society*, II.

Agar, N., 1981, *The Bedfordshire Farm Worker in the Nineteenth Century*, Bedfordshire Historical Record Society, CX.

Agricultural Economics Research Institute, Oxford, 1944, *Country Planning: A Study of Rural Problems*, Oxford University Press, Oxford.

Alexander, D., 1970, *Retailing in England during the Industrial Revolution*, Athlone Press, London.

Allen, C. W., 1914, 'The housing of the agricultural labourer', *Journal of the Royal Agricultural Society*, LXXV.

Ambrose, P., 1974, *The Quiet Revolution*. Chatto & Windus, London.

Anderson, M., 1971a, *Family Structure in Nineteenth Century Lancashire*, Cambridge University Press, Cambridge.

—— 1971b, 'Urban migration in nineteenth century Lancashire: some insights into two competing hypotheses', *Annales de Demographie Historique*, VIII.

—— 1976, 'Marriage patterns in Victorian Britain: an analysis based on registration district data for England and Wales, 1861', *Journal of Farming History*, I.

Arch, Joseph, 1898, (new edn 1966), *Joseph Arch: The Story of his Life, Told by Himself*, Hutchinson, London.

Armstrong, A., 1988, *Farmworkers. A Social and Economic History, 1770–1980*, Batsford, London.

Arnold, R., 1974, 'A Kentish exodus of 1879', *Cantium*, VI.

Aronson, H., 1914, *The Land and the Labourer*, Melrose, London.

Ashby, A., 1939, 'The effects of urban growth on the countryside', *Sociological Review*, XXXI.

Ashby, A. W. and Evans, I. L., 1944, *The Agriculture of Wales and Monmouthshire*, University of Wales Press, Cardiff.

Ashby, A. W. and Smith, J. H., 1938, 'Agricultural labour in Wales under statutory regulation of wages, 1924–37', *Welsh Journal of Agriculture*, XIV.

Ashby, J. and King, B., 1893, 'Statistics of some Midland villages', *Economic Journal*, III.

Ashby, M. K., 1933, 'Recent social changes as they affect the younger generation', *Journal of the Proceedings of the Agricultural Economics Society*, II.

Ashton, J. and Cracknell, B. E., 1961, 'Agricultural holdings and farm business structure in England and Wales', *Journal of Agricultural Economics*, XIV.

Atkinson, J. C., 1891, *Forty Years in a Moorland Parish*, Macmillan, London.

Ausubel, H., 1960, *In Hard Times: Reformers among the Late Victorians*, Columbia University Press, New York.

Baines, D. E., 1985, *Migration in a Mature Economy. Emigration and Internal Migration in England and Wales, 1861–1900*, Cambridge University Press, Cambridge.

Baker, M. 1974, *Folklore and Customs of Rural England*, David & Charles, Newton Abbot.

Bamford, T. W., 1965, *The Evolution of Rural Education, 1850–1964*, University of Hull Institute of Education, Hull.

Barley, M. W., 1961, *The English Farmhouse and Cottage*, Routledge & Kegan Paul, London.

Barnett, D. C., 1967, 'Allotments and the problem of rural poverty, 1780–1840', in E. L. Jones and G. E. Mingay (eds), *Land, Labour and Population in the Industrial Revolution*, Arnold, London.

Beeson, T., 1973, *The Church of England in Crisis*, Davis-Poynter, London.

Bell, Adrian, 1930, *Corduroy*, Cobden-Sanderson, London.

Bellerby, J. R., 1952, 'Comparison of skill, endurance and experience required in agriculture and industry', *Farm Economist*, VII.

—— 1953, 'Distribution of farm income in the U.K., 1867–1938', *Proceedings of the Agricultural Economics Society*, X.

—— 1956, *Agriculture and Industry Relative Income*, Macmillan, London.

—— 1968, 'The distribution of farm income in the U.K., 1867–1938', in W. E. Minchinton (ed.), *Essays in Agrarian History II*, David & Charles, Newton Abbot.

Bennett, E. N., 1914, *Problems of Village Life*, Williams & Norgate, London.

Bennett, L. G., 1943, 'The mobility of farm workers', *Farm Economist*, IV.

Bessell, J. E., 1972, *The Younger Worker in Agriculture: Projections to 1980*, NEDO, London.

Bielckus, C. L., Rogers, A. W., and Wibberley, G. P., 1972, *Second Homes in England and Wales*, University of London, Wye College, London.

Bienefeld, M. A., 1972, *Working Hours in British Industry*, Weidenfeld & Nicolson, London.

Bilson, C. J., 1895, *Country Folk-lore. Printed Extracts No. 3, Leicestershire and Rutland*, Folklore Society, London.

Black, M., 1968, 'Agricultural labour in an expanding economy', *Journal of Agricultural Economics*, XIX.

Board of Agriculture and Fisheries, 1913, *Report on Migration from Rural Districts in England and Wales*, HMSO, London.

Body, Richard, 1982, *Agriculture: the Triumph and the Shame*, Temple Smith, London.

Bonham-Carter, V., 1952, *The English Village*, Penguin, Harmondsworth.

Booth, Charles, 1894, *The Aged Poor in England and Wales*, Macmillan, London.

Bosanquet, B., 1950, 'The quality of the rural population', *Eugenics Review*, XLII.

Bowler, I. R., 1979, *Government and Agriculture*, Manchester University Press, Manchester.

Bracey, H. E., 1959, *English Rural Life: Village Activities, Organisations and Institutions*, Routledge & Kegan Paul, London.

Bradbury, J. B., 1981, *A History of Cockermouth*, Phillimore, Chichester.

Bradley, T. and Lowe, P. (eds), 1984, *Locality and Rurality: Economy and Society in Rural Regions*, Geo Books, Norwich.

Brigg Local History Group, 1983, *The Courts and Yards of Brigg*, Scunthorpe Museum Society, Scunthorpe, Lincolnshire.

Briggs, K. M., 1970, *A Dictionary of British Folk-tales in the English Language, Incorporating the F. J. Norton Collection: Part A, Folk Narratives*, Routledge & Kegan Paul, London.

—— 1971, *A Dictionary of British Folk-tales in the English Language, Part B, Folk Legends*, Routledge & Kegan Paul, London.

BPP (British Parliamentary Papers), 1867 XVII Registrar General of Births, Deaths and Marriages in England, 28th Annual Report.

—— 1867–8 XVII, XVIII R.C. Employment of Children, Young Persons and Women in Agriculture, First Report.

—— 1868–9 XIII R.C. Employment of Children, Young Persons and Women in Agriculture, Second Report.

—— 1882 XIV R.C. Depressed Condition of Agricultural Interests.

—— 1884–5 XXX R.C. Housing of the Working Classes, vol. II, Minutes of Evidence.

—— 1886 C.4848 Return of Number of Allotments and Agricultural Holdings in Great Britain.

—— 1890 C.6144 Return of Number of Allotments and Smallholdings.

—— 1893–4 XXXV R.C. Labour: The Agricultural Labourer (England).

—— 1893–4 XXXVI R.C. Labour: The Agricultural Labourer (Wales).

—— 1893–4 XXXVII pt II R.C. Labour: The Agricultural Labourer. General Report.

—— 1894 XVI pt 2 R.C. Agricultural Depression.

—— 1895 XVII R.C. Agricultural Depression.

—— 1900 LXIII Report by Wilson Fox on Wages and Earnings of Agricultural Labourers in U.K.

—— 1905 XCVIII Second Report of Wilson Fox on Wages and Earnings of Agricultural Labourers.

—— 1906 XCVI Board of Agriculture Report on Decline of Agricultural Population of Great Britain 1881–1906.

—— 1912–13 XI Registrar General of Births, Deaths and Marriages, Seventy-fourth Annual Report (for 1911).

Britton, D. K. and Ingersent, K. A., 1964, 'Trends in concentration of British agriculture', *Journal of Agricultural Economics*, XVI.

Britton, D. K. and Smith, J. H., 1947, 'Farm labour: problems of age composition and recruitment', *Farm Economist*, V.

Brown, J., 1986, *The English Market Town: A Social and Economic History*, Crowood Press, Ramsbury, Wilts.

Brown, M. and Winyard, S., 1975, *Low Pay on the Farm*, Low Pay Unit, Pamphlet no. 3, London.

Burnett, J., 1978, *A Social History of Housing 1815–1970*, Methuen, London.

Burton, H. M., 1953, *The Education of the Countryman*, Kegan Paul, London.

Caird, James, 1852, *English Agriculture in 1850–51*, Longman, London.

—— 1878, 'General view of British agriculture', *Journal of the Royal Agricultural Society*, 2nd ser., XIV.

—— 1968, (new edn) *English Agriculture in 1850–51*, Frank Cass, London.

Cairncross, A. J., 1953, *Home and Foreign Investment, 1870–1913*, Cambridge University Press, Cambridge.

Capstick, G. M., 1987, *Housing Dilemmas in the Lake District*, University of Lancaster Centre for North West Regional Studies, Manchester.

Carter, I. 1976, 'Class and culture among farm servants in the north-east', in A. A. Maclaren (ed.), *Social Class in Scotland*, John Donald, Edinburgh.
Caunce, Stephen, 1975, 'East Riding hiring fairs', *Oral History*, III, 2.
Cawte, E. C., Helm, A., and Peacock, N., 1967, *English Ritual Drama: A Geographical Index*, Folklore Society, London.
Chadwick, Edwin, 1842 (1965), *Report on the Sanitary Condition of the Labouring Population of Great Britain*, ed. M. W. Flinn, Edinburgh University Press, Edinburgh.
Chambers, J. D., 1972, *Population, Economy and Society in Pre-Industrial England*, Oxford University Press, Oxford.
Christian, Garth, 1966, *Tommorow's Countryside*, John Murray, London.
Clark, Colin, 1973, *The Value of Agricultural Land*, Pergamon, Oxford.
Clark, G. Kitson, 1973, *Churchmen and the Condition of England, 1823–1885*, Methuen, London.
Clifford, F., 1875, 'The Labour Bill in farming', *Journal of the Royal Agricultural Society*, 2nd ser., XI.
Cobbett, William, 1821, *Cottage Economy*, P. Davies, London.
Coleman, J., 1871, 'English cheese factories', *Country Gentleman's Magazine*, VI.
Collings, J., 1908, *Land Reform: Occupying Ownership, Peasant Proprietary and Rural Education*.
Collins, E. J. T., 1970, 'Harvest technology and labour supply 1790–1870', unpublished Ph.D. thesis, University of Nottingham, Nottingham.
—— 1972, 'The diffusion of the threshing machine in Britain 1780–1880', *Tools and Tillage*, II, 1.
—— 1976, 'Migrant labour in British agriculture in the nineteenth century', *Economic History Review*, 2nd ser., XXIX, 1.
Colls, R. and Dodds, P., 1986, *Englishness: Politics and Culture, 1880–1920*, Croom Helm, Beckenham.
Connell, J., 1974, 'The metropolitan village: spatial and social processes in "discontinuous suburbs" ', in J. H. Johnson (ed.), *Suburban Growth: Geographical Processes at the Edge of the City*, Wiley, Chichester.
—— 1978, *The End of Tradition: Country Life in Central Surrey*, Routledge & Kegan Paul, London.
Cooper, A. F., 1980, 'The transformation of agricultural policy, 1912–36', unpublished D. Phil. thesis, University of Oxford, Oxford.
Copper, B., 1971, *A Song for every Season*, Heinemann, London.
Coppock, J. T., 1971, *An Agricultural Geography of Great Britain*, Bell, London.
Cornish, J. G., 1939, *Reminiscences of Country Life*, Country Life, London.
Courtney, M. A., 1890, *Cornish Feasts and Folk-lore*, Beare & Son, Penzance.
Craigie, P. G., 1887, 'Size and distribution of agricultural holdings in England and abroad', *Journal of the Royal Statistical Society*, L.
Crichton, R., 1964, *Commuters' Village*, David & Charles, Dawlish.
Crosby, T. L., 1977, *English Farmers and the Politics of Protection*, Harvester Press, Hassocks, Sussex.

Dack, C., 1911, *Weather and Folklore of Peterborough and District* (reprinted (n.d.) from original edition of the Peterborough Natural History and Archaeological Society).
Davey, B. J., 1983, *Lawless and Immoral: Policing a Country Town 1838–1857*, Leicester University Press, Leicester.
David, P. A., 1970, 'Labour productivity in English agriculture, 1850–1914: some

quantitative evidence on regional differences', *Economic History Review*, 2nd ser., XXIII.

Davis, R. W., 1972, *Political Change and Continuity 1760–1885*, David & Charles, Newton Abbot.

Deane, P. and Cole, W. A., 1962 (2nd edn 1967), *British Economic Growth, 1688–1959*, Cambridge University Press, Cambridge.

Dent, J. D., 1871, 'The present condition of the English agricultural labourer', *Journal of the Royal Agricultural Society*, 2nd ser., VII.

Denton, J. B., 1864, *The Farm Homesteads of England*, William Clowes & Sons, London.

Dewey, P. E., 1975, 'Agricultural labour supply in England and Wales during the First World War', *Economic History Review*, 2nd ser., XXVIII.

—— 1979, 'Government provision of farm labour in England and Wales, 1914–18', *Agricultural History Review*, XXVII.

Digby, A., 1975, 'The labour market and the continuity of social policy after 1834: the case of the eastern counties', *Economic History Review*, 2nd ser., XXVIII.

Digby, M. and Gorst, S., 1957, *Agricultural Cooperation in the United Kingdom*, Blackwell, Oxford.

Dunbabin, J. P. D., 1974, *Rural Discontent in Nineteenth-Century Britain*, Faber, London.

Dyos, H. J. and Reader, D. A. 1972, 'Slums and suburbs', in H. J. Dyos and M. Wolff (eds), *The Victorian City, Images and Realities*, Routledge & Kegan Paul, London.

Dyos, H. J. and Wolff, M. (eds), 1972, *The Victorian City, Images and Realities*, Routledge & Kegan Paul, London.

Ellis, A. J., 1889, *On Early English Pronounciation, Part V: The Existing Phonology of English Dialects compared with that of West Saxon Speech*, Early English Text Society, London.

Emmett, I., 1964, *A North Wales Village: A Social Anthropological Study*, Routledge & Kegan Paul, London.

Erickson, C., 1976, *Emigration from Europe 1815–1914*, Black, London.

Evans, E. J., 1976, *The Contentious Tithe*, Routledge & Kegan Paul, London.

Evans, George Ewart, 1956, *Ask the Fellows Who Cut the Hay*, Faber, London.

—— 1960, 1967, *The Horse in the Furrow*, Faber, London.

—— 1966, *The Pattern under the Plough: Aspects of the Folk-life of East Anglia*, Faber, London.

—— 1970, *Where Beards Wag All: The Relevance of the Oral Tradition*, Faber, London.

—— 1975, *The Days that We have Seen*, Faber, London.

Evans, George Ewart and Thomson, D., 1972, *The Leaping Hare*, Faber, London.

Farrell, T., 1974, 'Report on the agriculture of Cumberland', *Journal of the Royal Agricultural Society*, 2nd ser., X.

Flux, A. W., 1930, 'Our food supply before and after the war', *Journal of the Royal Statistical Society*, XCIV.

Forsyth, D., 1984, The social effects of primary school closure', in T. Bradley and P. Lowe (eds), *Locality and Rurality*, Geo Books, Norwich.

Frankenberg, R., 1957, *Village on the Border: A Social Study of Religion, Politics and Football in a North Wales Community*, Cohen & West, London.

Gasson, R., 1966a, 'Part-time farmers in south-east England', *Farm Economist*, XI.

——1966b, 'The influence of urbanisation in farm ownership in practice, *Studies in Rural Land Use*, VII, University of London, Wye College, London.
—— 1969, *Occupational Immobility of Small Farmers*, University of Cambridge Department of Land Economy Occasional Paper XIII, Cambridge.
—— 1974a, *Mobility of Farm Workers*, University of Cambridge Department of Land Economy Occasional Paper II, Cambridge.
—— 1974b, 'Resources in agriculture: labour', in A. Edwards and A. J. Rogers (eds), *Agricultural Resources: An Introduction to the Farming Industry of the United Kingdom*, Faber, London.
—— 1975, *Provision of Tied Cottages*, University of Cambridge Department of Land Economy Occasional Paper IV, Cambridge.
Giles, A. K. and Cowie, W. J. G., 1964, *The Farm Worker: His Training Pay and Status*, University of Bristol Department of Agricultural Economics, Bristol.
Giles, A. K. and Mills, F. D., 1970, *Farm Managers Part 1*, University of Reading Department of Agricultural Economics Miscellaneous Studies XLVII, Reading.
Gilpin, M. C., 1960, 'Population changes round the shores of Milford Haven from 1800 to the present day', *Field Studies*, I.
Graham, P. Anderson, 1892, *The Rural Exodus: The Problem of the Village and the Town*, Methuen, London.
Green, F. E., 1920, *A History of the English Agricultural Labourer, 1870–1920*, P. S. King, London.
Greenhow, E. H., 1858 (reprinted 1973), *Papers Relating to the Sanitary State of the People of England*, Gregg International, London.
Grigg, D. B., 1963, 'A note on agricultural rent in nineteenth century England', *Agricultural History*, XXXVII.
Grigson, G., 1975, *The Englishman's Flora*, Paladin, St Albans.
Grinsell, L. V., 1976, *Folklore of Prehistoric Sites in Britain*, David & Charles, Newton Abbot.
Gurdon, Lady E. C., 1893, *Country Folklore, Printed Extracts No. 2, Suffolk*, Folklore Society, London.

Haggard, H. Rider, 1899 (2nd edn 1906a), *A Farmer's Year*, Longman, London.
—— 1902 (2nd edn 1906b), *Rural England*, Longman, London.
Haggard, L. (ed.), 1935, *I Walked by Night*, Nicholson & Watson, London.
Hallett, G., 1960, *The Economics of Agricultural Land Tenure*, Land Books, London.
Harrison, A., 1965, 'Some features of farm business structures', *Journal of Agricultural Economics*, XVI.
—— 1975, *Farmers and Farm Businesses in England*, University of Reading Department of Agricultural Economics and Management, Reading.
Harrison, E., 1928, *Harrison of Ightham*, Oxford University Press, Oxford.
Hartley, L. P., 1937, *The Go-Between*, Henry Hamilton, London.
Hartley, M. and Ingilby, J., 1968, *Life and Tradition in the Yorkshire Dales*, Dent, London.
Harvey, Nigel, 1970, *A History of Farm Buildings in England and Wales*, David & Charles, Newton Abbot.
Hasbach, W., 1966, *A History of the English Agricultural Labourer*, Frank Cass, London.
Havinden, M. A., 1966, *Estate Villages, A Study of the Berkshire Villages of Ardington and Lockinge*, Lund Humphreys, London.
Health and Safety Executive, 1982, *Health and Safety Statistics*, HMSO, London.
Heath, F. G., 1874, *The English Peasantry*, London.

Heath, R., 1893, *The English Peasant*, Unwin, London.
Henderson, William, 1866 (2nd edn 1879), *Notes on the Folk-lore of the Northern Counties of England and the Borders*, Folklore Society, London.
Hill, Francis, 1974, *Victorian Lincoln*, Cambridge University Press, Cambridge.
Hirsch, G. P., 1951, 'Migration from the land in England and Wales', *Farm Economist*, VI.
Holderness, B. A., 1972, 'Landlords' capital formation in East Anglia 1750–1870', *Economic History Review*, 2nd ser., XXV, 3.
—— 1985, *British Agriculture since 1945*, Manchester University Press, Manchester.
Hole, C., 1940 (2nd edn 1944–5), *English Folklore*, Batsford, London.
Holland, W., 1984, *Paupers and Pig Killers: The Diary of William Holland, A Somerset Parson, 1799–1828*, ed. J. Ayres, Alan Sutton, Gloucester.
*Hop Pocket, The* (mimeographed newsletter series, held at NUAAW office, Maidstone).
Horn, Pamela, 1972, 'Agricultural trade unionism and emigration, 1872–1881', *Historical Journal*, XV, 1.
—— 1974, *The Victorian Country Child*, Roundwood Press, Kineton, Warwickshire.
—— 1976, *Labouring Life in the Victorian Countryside*, Gill & Macmillan, Dublin.
Howell, D. W., 1978, *Land and People in Nineteenth Century Wales*, Routledge & Kegan Paul, London.
Howitt, W., 1838 (3rd edn 1971), *The Rural Life of England*, Irish University Press, Dublin.
Howkins, A., 1973, *Whitsun in Nineteenth Century Oxfordshire*, History Workshop Pamphlets No. 8, Ruskin College, Oxford.
—— 1985, *Poor Labouring Men. Rural Radicalism in Norfolk, 1870–1923*, History Workshop Series, Routledge & Kegan Paul, London.
—— 1986, 'The rediscovery of rural England', in R. Colls and P. Dodds (eds), *Englishness: Politics and Culture 1880–1920*, Croom Helm, Beckenham.
Hughes, J. D., 1957, 'A note on the decline in numbers of farm workers in Great Britain', *Farm Economist*, VIII.
Hunt, E. H., 1967, 'Labour productivity in English agriculture, 1850–1914', *Economic History Review*, 2nd ser., XX.
—— 1973, *Regional Wage Variations in Britain, 1850–1914*, Clarendon Press, Oxford.
Hurt, J., 1961, 'The role of the Hertfordshire gentry and the education committee of the Privy Council in providing education in Hertfordshire in the nineteenth century', unpublished Ph.D. thesis, University of London, London.
—— 1968, 'Landowners, farmers and clergy in the financing of rural education before 1870', *Journal of Education Administration and History*, I, 1.
Hutt, C., 1973, *The Death of the English Pub*, Arrow Books, London.
Huzel, James, P., 1980, 'The demographic impact of the Old Poor Law: more reflexions on Malthus', *Economic History Review*, 2nd ser., XXXIII,3.

Innes, J. W., 1938, *Class Fertility Trends in England and Wales, 1876–1934*, Princeton University Press, Princeton, NJ.

Jefferies, Richard, 1880, *Hodge and his Masters*, Smith, Elder, London.
Jenkins, D., 1971, *The Agricultural Community in South-West Wales at the turn of the Twentieth Century*, University of Wales Press, Cardiff.
Jenkins, J. G., 1976, *Life and Tradition in Rural Wales*, Dent, London.
Jennings, P., 1965, *The Living Village*, Hodder & Stoughton, London.
Jessopp, Augustus, 1887, *Arcady, for Better, for Worse*, T. Fisher Unwin, London.

Johnson, J. H., 1974, *Suburban Growth: Geographical Processes at the Edge of the City*, Wiley, Chichester.
Johnston, H. J. M., 1972, *British Emigration Policy, 1815–1830: Shovelling out Paupers*, Clarendon Press, Oxford.
Johnston, W., 1851, *England As It Is . . . in the Middle of the Nineteenth Century.*
Jones, E. L., 1964, 'The agricultural labour market in England, 1793–1872', *Economic History Review*, 2nd ser., XVII.
Jones, G. P., 1962, 'The decline of the yeomanry in the Lake Counties', *Transactions of the Cumberland and Westmorland Antiquarian Society*, LXII.
Jones, R. E., 1968, 'Population and agrarian change in an eighteenth-century Shropshire parish', *Local Population Studies*, I.
—— 1976, 'Infant mortality in rural north Shropshire, 1561–1810', *Population Studies*, XXX.
Jones-Baker, D., 1977, *The Folklore of Hertfordshire*, Batsford, London.
Joyce, P., 1980, *Work, Society and Politics: The Culture of the Factory in Later Victorian England*, Harvester Press, Hassocks, Sussex.

Keatinge, G. F. and Littlewood, R., 1948, 'The agricultural worker', *The Lancet*, 21 August.
Kerr, B., 1968, *Bound to the Soil*, John Baker, London.
Killip, M., 1975, *The Folklore of the Isle of Man*, Batsford, London.
Kitchen, F., 1940, *Brother to the Ox*, Dent, London.

Land Enquiry Committee, 1913, *The Land, vol. I, Rural*, Hodder & Stoughton, London.
Lawton, R., 1973, 'Rural depopulation in nineteenth century England', in D. R. Mills (ed.), *English Rural Communities: The Impact of a Specialized Economy*, Macmillan, London.
Leather, E. M., 1970, *The Folk-lore of Herefordshire*, S. R. Publishing, Wakefield.
Lennard, R., 1914, *Economic Notes on English Agricultural Wages*, Macmillan, London.
Liberal Land Committee, 1925, *The Land and the Nation*, Hodder & Stoughton, London.
Little, E., 1845, 'The farming of Wiltshire', *Journal of the Royal Agricultural Society*, V.
Littlejohn, J., 1963, *Westrigg: The Sociology of a Cheviot Parish*, Routledge & Kegan Paul, London.
Llewellyn Smith, H., 1904, 'The influx of population', in Charles Booth (ed.), *Life and Labour of the People in London*, III, pt 1.
Lorrain Smith, E., 1932, *Go East for a Farm: A Study of Rural Migration*, University of Oxford Institute for Research in Agricultural Economics, Oxford.
Lucey, B., 1978, *Twenty Centuries in Sedlescombe: An East Sussex Parish*, Regency Press, London.
Lund, P. J., Morris, T. G., Temple, J. D., and Watson, J. M., 1982, *Wages and Employment in Agriculture: England and Wales, 1960–1980*, Ministry of Agriculture, Fisheries and Food, Government Economic Service Working Paper no. 52, HMSO, London.

McConnell, P., 1891, 'Experiences of a Scotsman on the Essex clays', *Journal of the Royal Agricultural Society*, 3rd ser., II.
Madden, M., 1956, 'The National Union of Agricultural Workers, 1906–1956', unpublished B. Litt.thesis, University of Oxford, Oxford.

Malcolmson, R. W., 1973, *Popular Recreation in English Society 1700–1850*, Cambridge University Press, Cambridge.
—— 1981, 'Leisure', in G. E. Mingay (ed.), *The Victorian Countryside*, Routledge & Kegan Paul, London.
Martin, A. F. and Steel, R. W. (eds), 1954, *The Oxford Region*, Oxford University Press, Oxford.
Martin, J. M., 1976, *The Rise in Population in Eighteenth Century Warwickshire*, Dugdale Society Occasional Papers, no. 23, Oxford.
Mechi, J. J., 1845, *Letters on Agricultural Improvement*, London.
—— 1859, *How to Farm Profitably: Or the Sayings and Doings of Mr Alderman Mechi*, London.
Mejer, E., 1949, *Agricultural Labour in England and Wales, I, 1900–20*, University of Nottingham Department of Agricultural Economics, Nottingham.
—— 1951, *Agricultural Labour in England and Wales, II, 1917–51*, University of Nottingham Department of Agricultural Economics, Nottingham.
Melling, E., 1964, *Kentish Sources: IV, The Poor*, Kent County Archives Office, Maidstone.
Metcalf, D., 1969, 'Labour productivity in English agriculture 1850–1914', *Economic History Review*, 2nd ser., XXII.
Miller, C., 1984, 'The hidden workforce: female fieldworkers in Gloucestershire, 1870–1901', *Southern History*, VI.
Minchinton, W., 1975, 'Cider and folklore', *Folk Life*, XIII.
Ministry of Agriculture, Fisheries and Food, 1967, *The Changing Structure of the Agricultural Labour Force in England and Wales. Numbers of Workers, Hours and Earnings, 1945–65*, HMSO, London.
—— 1968, *A Century of Agricultural Statistics*, HMSO, London.
Mingay, G. E. (ed.), 1981, *The Victorian Countryside*, Routledge & Kegan Paul, London.
'Miss Read', 1960, *Village School*, Penguin, Harmondsworth.
Mollett, J. A., 1949, 'An economic study of the supply of labour in Buckinghamshire', unpublished M.Sc. thesis, University of Reading.
Moreau, R. E., 1968, *The Departed Village: Barrick Salome at the Turn of the Century*, Oxford University Press, Oxford.
Morgan, David, 1975, 'The place of harvesters in nineteenth-century village life', in R. Samuel (ed.), *Village Life and Labour*, Routledge & Kegan Paul, London.
—— 1982, *Harvests and Harvesting 1840–1900*, Croom Helm, Beckenham.
Morton, J. C., 1861 (new edn 1868), *Handbook of Farm Labour*, Cassell, Petter & Galpin, London.
Moseley, M. J. (ed.), 1978, *Social Issues in Rural Norfolk*, University of East Anglia Centre for East Anglian Studies, Norwich.
Murray, K. A. H., 1955, *Agriculture, History of the Second World War, United Kingdom, Civil Series*, HMSO, London.

Nalson, J. S., 1968, *The Mobility of Farm Families*, Manchester University Press, Manchester.
Nash, R., 1980, *Schooling in Rural Societies*, Methuen, London.
Newby, H., 1977, *The Deferential Worker*, Methuen, London.
—— 1979, *Green and Pleasant Land? Social Change in Rural England*, Hutchinson, London.
Newby, H. (ed.), 1978, *International Perspectives in Rural Sociology*, Wiley, Chichester.

Nicholls, G., 1846, 'On the condition of the agricultural labourer', *Journal of the Royal Agricultural Society*, VII.
Northall, G. F., 1892, *English Folk-Rhymes*, Kegan Paul, London.

Obelkevitch, J., 1976, *Religion and Rural Society: South Lindsey 1825–1875*, Clarendon Press, Oxford.
Office of Population Censuses and Surveys, 1975, *Census, 1971, England and Wales. Economic Activity. County Leaflet, Norfolk*, HMSO, London.
Ogle, W., 1889, 'The alleged depopulation of the rural districts of England', *Journal of the Statistical Society*, LII.
Olive, G. W., 1938, 'Education in rural schools', *Journal of the Farmers' Club*, pt 2, March.
Olney, R. J., 1973, *Lincolnshire Politics 1832–1885*, Oxford University Press, Oxford.
Opie, I. and Opie, P., 1959, *The Lore and Language of Schoolchildren*, Oxford University Press, Oxford.
Orwin, C. S., 1945, *The Problems of Rural Life*, Oxford.
Orwin, C. S. and Felton, B. I., 1931, 'A century of wages and earnings in agriculture', *Journal of the Royal Agricultural Society*, XCII.
Orwin, C. S. and Whetham, E. H., 1964, *History of British Agriculture, 1846–1914*, Longman, London.

Pahl, R., 1964, Urbs in Rure: *The Metropolital Fringe in Hertfordshire*, London School of Economics Geographical Paper, London.
—— 1966, 'The rural-urban continuum', *Sociologia Ruralis*, VI.
Paul, L., 1964, *The Deployment and Payment of the Clergy*, Church Information Office, London.
Pedley, W. H., 1942, *Labour on the Land*, King & Staples, London.
Perkins, J. A., 1975, 'Tenure, tenant right and agricultural progress in Lindsey, 1780–1850', *Agricultural History Review*, XXIII, 1.
Perren, Richard, 1970, 'The landlord and agricultural transformation, 1870–1900', *Agricultural History Review*, XVIII, 1.
Peters, J. E. C., 1969, *The Development of Farm Buildings in the Western Lowlands of Staffordshire up to 1885*, Manchester University Press, Manchester.
Phillips, A. D. M., 1969, 'Underdraining and the English claylands, 1850–80: a review', *Agricultural History Review*, XVII, 1.
Phythian-Adams, C., 1975, *Local History and Folklore: A New Framework*, National Council of Social Services Standing Conference for Local History, London.
Plomer, W. (ed.), 1964, *Kilvert's Diary 1870–1879: Selections from the Diary of the Rev. Francis Kilvert*, Jonathan Cape, London.
Poole, R., 1985, 'Walks, holidays and pleasure fairs in the Lancashire cotton district, 1790–1890', unpublished Ph.D. thesis, University of Lancaster, Lancaster.
Pratt, E. A., 1906, *The Transition in Agriculture*, John Murray, London.

Raistrick, A., 1968, *The Pennine Dales*, Eyre & Spottiswoode, London.
Ratcliffe, H., 1850, *Observations on the Rate of Mortality and Sickness Existing Among Friendly Societies: particularised for various trades, occupations and localities. With a series of tables showing the value of annuities, sick-gift, assurance for death. Calculated from the experience of the Members Company, the Independent Order of Odd Fellows, Manchester Unity Friendly Society*, Manchester.

Ravenstein, E. G., 1885, 'The laws of migration', *Journal of the Statistical Society*, XLVIII.
Redford, A., 1926 (new edn 1964), *Labour Migration in England, 1800–1850*, Manchester University Press, Manchester.
Ree, H., 1973, *Educator Extraordinary: The Life and Achievements of Henry Morris, 1889–1961*, Longman, London.
Roberts, R., 1971 (reprinted 1973), *The Classic Slum*, Penguin, Harmondsworth.
Robertson Scott, J. W., 1926, *The Dying Peasant*, Williams & Norgate, London.
Robin, J., 1980, *Elmdon. Continuity and Change in a North-West Essex Village, 1861–1964*, Cambridge University Press, Cambridge.
Rogers, A., 1989, 'A planned countryside', in G. E. Mingay (ed.), *The Rural Idyll*, Routledge, London.
Rollinson, W., 1974, *Life and Tradition in the Lake District*, Dent, London.
Rolls, M. J., 1965, 'Some aspects of rural education in England and Wales', *Comparative Education Review*, IX.
Rowntree, B. Seebohm and Kendall, May, 1913, *How the Labourer Lives*, Nelson, London.
Ruddock, E., 1964–5, 'May-day songs and celebrations in Leicestershire and Rutland', *Transactions of the Leicestershire Archaeological and Historical Society*, XL.
Russell, A., 1986, *The Country Parish*, SPCK, London
Russell, Rex C., 1965–7, *History of Schools and Education in Lindsey 1800–1902*, Lindsey County Council, Lincoln.

Samuel, Raphael, 1972, 'Comers and goers', in H. J. Dyos and M. Wolff (eds), *The Victorian City*, Routledge & Kegan Paul, London.
——1975, *Village Life and Labour*, Routledge & Kegan Paul, London.
Saville, J., 1957, *Rural Depopulation in England and Wales 1851–1951*, Routledge & Kegan Paul, London.
Self, P. and Storing, H. J., 1962, *The State and the Farmer*, Allen & Unwin, London.
Sellman, R. R., 1967, *Devon Village Schools in the Nineteenth Century*, David & Charles, Newton Abbot.
Seward, W. R., 1937, 'Technical instruction and the agricultural worker', *Journal of the Farmers' Club*, Part 3, April.
Shaw, J. M., 1978, 'The social implications of village development', in M. J. Moseley (ed.), *Social Issues in Rural Norfolk*, University of East Anglia Centre for East Anglian Studies, Norwich.
Shears, R. T., 1936, 'Housing the agricultural worker', *Journal of the Royal Agricultural Society*, 3rd ser., XCVII.
Shoard, M., 1980, *The Theft of the Countryside*, Temple Smith, London.
Shorter, A. H., Ravenhill, W. L. D., and Gregory, K. F., 1969, *Southwest England*, Nelson, London.
Sidney, Samuel, 1848, *Railways and Agriculture in North Lincolnshire.*
Simpson, J., 1973, *The Folklore of Sussex*, Batsford, London.
Snell, K. D. M., 1985, *Annals of the Labouring Poor. Social Change and Agrarian England, 1660–1900*, Cambridge University Press, Cambridge.
*Social Trends*, 1987, HMSO, London.
Stacey, M., 1969, 'The myth of community studies', *British Journal of Sociology*, IV.
Standfield, F. G., 1984, *A History of East Meon*, Phillimore, Chichester.

Stebbing, S., 1984, 'Women's roles and rural society', in T. Bradley and P. Lowe (eds), *Locality and Rurality*, Geo Books, Norwich.
Stedman Jones, G., 1971, *Outcast London: A Study in the Relationship between Classes in Victorian Society*, Clarendon Press, Oxford.
Steel, D. I. A., 1979, *A Lincolnshire Village. The Parish of Corby Glen in its Historical Context*, Longman, London.
Street, A. G., 1932, *Farmer's Glory*, Faber, London.
Sturmey, S. G., 1968, 'Owner-farming in England and Wales, 1900–50', in W. E. Minchinton (ed.), *Essays in Agrarian History, II,* David & Charles, Newton Abbot.
Sturt, G. (George Bourne), 1912, *Change in the Village*, Duckworth, London.
Sutherland, D., 1968, *The Landowners*, Blond, London.

Taylor, F. D. W., 1955, 'United Kingdom: numbers in agriculture', *Farm Economist*, VIII.
Thackrah, C. T., 1832 (reprinted 1957), *The Effects of Arts, Trades and Professions on Health and Longevity, with suggestions for the removal of many of the agents which produce disease and shorten the duration of life*, E. & S. Livingstone, Edinburgh.
Thirsk, Joan, 1957, *English Peasant Farming*, Routledge & Kegan Paul, London.
Thompson, Flora, 1939 (reprinted 1954), *Lark Rise to Candleford*, Oxford University Press, Oxford.
Thompson, F. M. L., 1963, *English Landed Society in the Nineteenth Century*, Routledge & Kegan Paul, London.
Thompson, R. J., 1968, 'An enquiry into the rent of agricultural land', in W. E. Minchinton (ed.), *Essays in Agrarian History, II,* David & Charles, Newton Abbot.
Tonnies, F., 1887, *Gemeinschaft und Gesellschaft*, English edn, *Community and Society*, 1957, Harper Torchbooks, New York.
Torr, C., 1918, *Small Talk at Wreyland,* Cambridge University Press, Cambridge.
Tranter, N. L., 1985, *Population and Society, 1750–1940*, Longman, London.

*Victoria County History: Staffordshire VI*, 1979, eds M. W. Greenslade and D. A. Johnson, Oxford University Press, Oxford.
*Victoria County History: Sussex VI, 1*, 1980, ed. T. P. Hudson, Oxford University Press, Oxford.

Wailes, R., 1954, *The English Windmill*, Routledge & Kegan Paul, London.
Wakelin, M. F., 1972, *English Dialects: An Introduction*, Athlone Press, London.
Ward, W. R., 1965, 'The tithe question in England in the early nineteenth century', *Journal of Ecclesiastical History*, XVI, 1.
Whetham, E. H., 1968, 'Sectoral advance in English agriculture, 1850–80: a summary', *Agricultural History Review*, XVI.
Whitby, M. C., 1966, 'Farmers in England and Wales, 1921–61', *Farm Economist*, XI.
——1967, 'Labour mobility and training in agriculture', *Westminster Bank Review*, August.
White, Arnold, 1901, *Efficiency and Empire*, Methuen, London.
Wymer, N. G., 1948, *English Country Crafts*, Batsford, London.
Williams, H. T. (ed.), 1960, *Principles for British Agricultural Policy*, Oxford University Press, Oxford.
Williams, R., 1973, *The Country and the City*, Chatto & Windus, London.
Williams, W. M., 1956, *The Sociology of an English Village: Gosforth*, Routledge, London.

——1958, *The Country Craftsman: A Study of Some Rural Crafts and the Rural Industries Organisation in England,* Routledge & Kegan Paul, London.
——1963, *A West Country Village: Ashworthy, Pt. 1,* Routledge & Kegan Paul, London.
Williamson, Henry, 1941, *The Story of a Norfolk Farm*, Faber, London.
Wilson, J. M., 1851, *The Rural Cyclopaedia*, Edinburgh.
Wise, M., 1931, *English Village Schools*, Hogarth Press, London.
Woodruff, D., 1934, 'Expansion and emmigration', in G. M. Young (ed.), *Early Victorian England, 1830–1865*, Oxford University Press, Oxford.
Woods, K. S., 1949 (new edn 1975), *Rural Crafts of England*, E. P. Publishing, East Ardsley, Wakefield.
Wormell, Peter, 1978, *Anatomy of Agriculture: A Study of Britain's Greatest Industry*, Harrap, London.
Wrigley, E. A., 1985, 'Urban growth and agricultural change: England and the continent in the early modern period', *Journal of Interdisciplinary History*, XV.
——1986, 'Men on the land and men in the countryside: employment in agriculture in early nineteenth century England', in L. Bonfield, R. M. Smith, and K. Wrightson (eds), *The World We Have Gained. Histories of Population and Social Structures,* Blackwell, Oxford.
Wrigley, E. A. and Schofield, M. S., 1981, *The Population History of England, 1541–1821*, Edward Arnold, London.
Wrigley, J., 1946, 'The rising costs of labour on fifteen farms in the eastern counties, 1940–1945', *Farm Economist*, V.

Young, Arthur, 1804, *General View of the Agriculture of Norfolk*, Board of Agriculture, London.
Young, Rev. Arthur, 1813, *General View of the Agriculture of Sussex*, Board of Agriculture, London.
Young, M. and Willmott, P., 1957, *Family and Kinship in East London*, Routledge & Kegan Paul, London.

# Index